VOLUME ONE HUNDRED AND SIXTY FOUR

CURRENT TOPICS IN DEVELOPMENTAL BIOLOGY

Cell Migration in Development

CURRENT TOPICS IN DEVELOPMENTAL BIOLOGY

"A meeting-ground for critical review and discussion of developmental processes"
A.A. Moscona and Alberto Monroy (Volume 1, 1966)

VOLUME ONE HUNDRED AND SIXTY FOUR

CURRENT TOPICS IN DEVELOPMENTAL BIOLOGY

Cell Migration in Development

Edited by

ROBERTO MAYOR

Department of Cell & Developmental Biology,
University College London, Gower Street,
London, WC1E 6BT, United Kingdom

ELSEVIER

ACADEMIC PRESS
An imprint of Elsevier

Academic Press is an imprint of Elsevier
125 London Wall, London, EC2Y 5AS, United Kingdom
50 Hampshire Street, 5th Floor, Cambridge, MA 02139, United States

First edition 2025

ISBN: 978-0-323-99926-7
ISSN: 0070-2153

For information on all Academic Press publications
visit our website at https://www.elsevier.com/books-and-journals

Publisher: Zoe Kruze
Editorial Project Manager: Devwart Chauhan
Production Project Manager: Maria Shalini
Cover Designer: Arumugam Kothandan

Typeset by MPS Limited, India

Contents

Contributors *ix*
Preface *xi*

1. Single-cell migration in development – Lessons from germ cells **1**

Jan Schick, Laura Ermlich, Lucas Kühl, Dennis Hoffmann, and Erez Raz

1. Overview 2
2. Motility of PGCs – principles and mechanisms 3
3. Tissue barriers in PGC migration 8
4. Directing PGC migration 11
5. Increasing migration precision 16
6. Perspectives 19

Acknowledgments 21
References 21

2. Outside influences: The impact of extracellular matrix mechanics on cell migration **29**

Ronen Zaidel-Bar and Priti Agarwal

1. Introduction 30
2. ECM composition and structure 30
   2.1 Interstitial matrix 31
   2.2 Basement membrane 33
3. Modulators of ECM mechanics 34
   3.1 Cross-linking enzymes 35
   3.2 Matrix proteases 35
4. Mechanical role of ECM in cell migration 36
   4.1 ECM fibrillar architecture 36
   4.2 ECM porosity 42
   4.3 ECM dynamics 45
   4.4 Cell-matrix anchorage 47
   4.5 ECM Stiffness 49
5. Outlook 52

References 55

3. From neural crest migration to the onset of gangliogenesis **67**

Hugo A. Urrutia and Marianne E. Bronner

1. Introduction 68
2. Neural crest induction and formation 72
3. Neural crest EMT and onset of migration 74
4. Cranial neural crest migration 75
   4.1 Trunk neural crest migration 80
5. Onset of cranial neural crest ectomesenchymal differentiation 82
   5.1 Cranial neural crest and placode contribution to peripheral nervous systems 84
   5.2 Epibranchial ganglia 87
   5.3 Trigeminal ganglia 88
6. Concluding remarks 91

Acknowledgments 92

References 92

4. Collective migration in complex environments: The case of the trunk neural crest **109**

Hanna-Maria Häkkinen, Soraya Villaseca, Marie De La Burgade, Zainalabdeen Alhashem, and Elena Scarpa

1. Introduction 110
2. Brief history of trunk neural crest research 112
3. The trunk and cranial neural crest migratory pathways 113
   3.1 Cranial neural crest migratory pathway 113
4. Trunk neural crest migratory pathways 114
   4.1 The ventromedial pathway 115
   4.2 The dorsolateral pathway 115
5. Cell fates of the trunk neural crest 116
   5.1 Models of fate acquisition in the TNC 117
   5.2 Mechanisms of fate acquisition in the zebrafish ventromedial TNC streams 118
   5.3 A role for TNC spatial context for cell fate specification? 120
6. Modes of neural crest migration 121
   6.1 Collective cell migration in NCCs 121
   6.2 Collective and individual cell migration in TNCs 124
7. Patterning of trunk NC migration and interaction with surrounding environment 127
   7.1 Trunk neural crest interactions with somites 128
   7.2 Interactions with the extracellular matrix 131

7.3 Physical cues and NC migration 134
8. In vitro models of neural crest: advantages and limitations 135
9. Conclusion 138
References 139

5. Cell migration during development: Is it always collective and never as a single cell? **153**

Jorge Diaz, Amrutha Patkunarajah, and Roberto Mayor

1. Introduction 154
2. Amoeboid and mesenchymal modes of cell migration 156
3. Single cell migration in development 159
3.1 Primordial germ cell migration 159
3.2 Zebrafish primordial germ cell migration 160
3.3 *Drosophila melanogaster* primordial germ cell migration 163
4. Collective cell migration: epithelial and mesenchymal 166
5. Collective cell migration in development 168
5.1 Lateral line in zebrafish 168
5.2 Drosophila border cells 169
5.3 Drosophila testis myotubes 173
6. Beyond collective cell migration: supracellular organization 174
6.1 Neural crest migration in *Xenopus* 175
6.2 Drosophila melanogaster gastrulation 176
7. Concluding remarks 177
Acknowledgments 179
References 179

Contributors

Priti Agarwal
Department of Cell and Developmental Biology, School of Medicine, Faculty of Medical and Health Sciences, Tel Aviv University, Tel Aviv, Israel; National Center for Biological Sciences, Tata Institute for Fundamental Research, Bangalore, Karnataka, India

Zainalabdeen Alhashem
Department of Physiology, Development and Neuroscience, University of Cambridge, Cambridge, United Kingdom

Marianne E. Bronner
Division of Biology and Biological Engineering, California Institute of Technology, Pasadena, CA, United States

Marie De La Burgade
Department of Physiology, Development and Neuroscience, University of Cambridge, Cambridge, United Kingdom

Jorge Diaz
Department of Cell and Developmental Biology, Anatomy Building, University College London, London, United Kingdom

Laura Ermlich
Institute of Cell Biology, Center for Molecular Biology of Inflammation, University of Münster, Münster, Germany

Dennis Hoffmann
Institute of Cell Biology, Center for Molecular Biology of Inflammation, University of Münster, Münster, Germany

Hanna-Maria Häkkinen
Department of Physiology, Development and Neuroscience, University of Cambridge, Cambridge, United Kingdom

Lucas Kühl
Institute of Cell Biology, Center for Molecular Biology of Inflammation, University of Münster, Münster, Germany

Roberto Mayor
Department of Cell and Developmental Biology, Anatomy Building, University College London, London, United Kingdom

Amrutha Patkunarajah
Department of Cell and Developmental Biology, Anatomy Building, University College London, London, United Kingdom

Erez Raz
Institute of Cell Biology, Center for Molecular Biology of Inflammation, University of Münster; Max Planck Institute for Molecular Biomedicine, Münster, Germany

Elena Scarpa
Department of Physiology, Development and Neuroscience, University of Cambridge, Cambridge, United Kingdom

Jan Schick
Institute of Cell Biology, Center for Molecular Biology of Inflammation, University of Münster, Münster, Germany

Hugo A. Urrutia
Division of Biology and Biological Engineering, California Institute of Technology, Pasadena, CA, United States

Soraya Villaseca
Department of Physiology, Development and Neuroscience, University of Cambridge, Cambridge, United Kingdom

Ronen Zaidel-Bar
Department of Cell and Developmental Biology, School of Medicine, Faculty of Medical and Health Sciences, Tel Aviv University, Tel Aviv, Israel

Preface

Cell migration is a fundamental biological process that plays a crucial role in embryonic development, tissue regeneration, and disease progression. Understanding the mechanisms that drive single-cell and collective migration is essential for deciphering how complex biological structures form and function. This special issue of **Current Topic on Developmental Biology** titled "*Cell Migration in Development*" brings together a collection of chapters that explore key aspects of cell migration, with a focus on developmental contexts and the underlying cellular and molecular mechanisms.

In the opening chapter, *Single-cell migration in development - lessons from germ cells*, Erez Raz lab explores the migration of germ cells, a unique model for understanding directed single-cell movement. Germ cells exhibit remarkable precision in their journey towards the gonads, guided by intricate signaling cues. Their study provides valuable insights into the broader principles of single-cell migration.

Shifting focus to the extracellular environment, Ronen Zaidel-Bar and and Priti Agarwal discuss the link between *Extracellular matrix mechanics and cell migration*. This chapter delves into the dynamic interactions between migrating cells and the extracellular matrix, highlighting how changes in adhesion and matrix properties influence collective movements.

Marianne Bronner and Hugo A. Urrutia, in their chapter *From neural crest migration to the onset of gangliogenesiss*, examine the transition of neural crest cells from a migratory state to differentiation and integration into functional structures. This crucial step in development underscores the fine balance between motility and differentiation that defines successful gangliogenesis.

The theme of neural crest migration continues with Elena Scarpa's contribution, *Collective migration in complex environments: the case of trunk neural crest*. She provides a detailed exploration of how neural crest cells navigate constrained environments, shedding light on the unique biomechanical and signaling adaptations that enable efficient migration in confined spaces.

Finally, Jorge Diaz, Amrutha Patkunarajah, and Roberto Mayor present *From single to collective: the mechanisms underlying cell migration in early development*, offering a comprehensive synthesis of the transition from individual to collective migration. Their work bridges the gap between single-cell behavior and coordinated tissue movement, emphasizing the emergent properties of collective migration in embryonic development.

Together, these chapters provide a multifaceted perspective on cell migration, integrating insights from different model systems and levels of biological organization. We hope this book serves as a valuable resource for researchers, students, and anyone interested in the fascinating world of cell movement and its implications for development and disease.

Roberto Mayor
Department of Cell & Developmental Biology,
University College London, Gower Street,
London, WC1E 6BT, United Kingdom

CHAPTER ONE

Single-cell migration in development – Lessons from germ cells

Jan Schick[a], Laura Ermlich[a], Lucas Kühl[a], Dennis Hoffmann[a], and Erez Raz[a,b,*]
[a]Institute of Cell Biology, Center for Molecular Biology of Inflammation, University of Münster, Münster, Germany
[b]Max Planck Institute for Molecular Biomedicine, Münster, Germany
*Corresponding author. e-mail address: erez.raz@uni-muenster.de

Contents

1. Overview 2
2. Motility of PGCs – principles and mechanisms 3
3. Tissue barriers in PGC migration 8
4. Directing PGC migration 11
5. Increasing migration precision 16
6. Perspectives 19
Acknowledgments 21
References 21

Abstract

Primordial germ cells (PGCs) migrate to associate with somatic cells to form the gonad, generate gametes and thereby ensure fertility. This chapter presents the mechanisms underlying single-cell migration performed by PGCs in various organisms, such as zebrafish, Drosophila, and the mouse models. This review introduces the principles of cell motility, factors controlling directed migration, and the effects of interactions between migrating cells and their environment. Specifically, it discusses passive and active migration mechanisms, the roles of guidance cues, and of interactions with different tissues that influence PGC migration. Comparative analysis of the process in different organisms reveals conserved and distinct strategies for motility and directed migration. The presented mechanisms contribute to broader understanding of cell migration, highlighting PGCs as a useful in vivo model for studying the principles governing the movement of cells within tissues.

Current Topics in Developmental Biology, Volume 164
ISSN 0070-2153, https://doi.org/10.1016/bs.ctdb.2025.01.002

1. Overview

In many animals, particularly in the common models used in developmental biology, primordial germ cells (PGCs) are specified at very early embryonic stages and become committed to the germline lineage to ensure the fertility and transmission of the genetic information to the next generation (Extavour & Akam, 2003). During development, PGCs associate with a specific set of somatic cells to form the gonad, the organ where the gametes, sperm and egg are produced. Because germ cells and the somatic cells that support their development are generated at different time points, embryogenesis is constrained, specifically by the need for proximity between the cell lineages, which restricts the locations where these lineages are specified and further differentiated. Indeed, PGCs address this constraint via their ability to migrate as single cells over long distances and interact with somatic cells, which supports their further development (Grimaldi & Raz, 2020; Richardson & Lehmann, 2010). Notably, this property of PGCs allows more freedom in the organism's embryonic development by allowing the specification of lineages that give rise to somatic-gonad cells and those that give rise to the germline to take place at distant locations. Thus, because PGCs can migrate, organisms can change their developmental trajectories and morphology during evolution, a concept termed "evolvability" (Grimaldi & Raz, 2020; Kirschner & Gerhart, 1998). In addition, to successfully migrate and arrive at the region where the gonad develops, germ cells must resist the differentiation cues they encounter on the way and maintain their fate. Failure to do so can affect the development of PGCs, as they may turn into other cell types via differentiation and can form germ cell tumors (Gross-Thebing et al., 2017; Oosterhuis & Looijenga, 2019; Westerich et al., 2023).

Overall, studying the migration of PGCs represents an excellent model for investigating the mechanisms that govern the acquisition of motility, the factors that allow for sustained directional migration, the interactions between migrating cells and their environment, and the basis for their ability to perceive and respond to guidance cues. PGC migration is being studied in different model organisms; in *Drosophila* in particular, genetic screens have led to the identification of many genes that have important functions for PGC migration (Moore, Broihier, Doren, Lunsford, & Lehmann, 1998). Further, studies in zebrafish have led to the discovery of vertebrate-specific proteins that play a role in migration (e.g., chemokines and their receptors) (Doitsidou et al., 2002; Richardson & Lehmann, 2010).

The following sections present the findings and insights derived from studies on the migration of PGCs in in vivo contexts across embryos of various species. As a model organism for studying germ cell migration, the zebrafish offers unique features such as external development and optical clarity of the embryo. Thus, in addition to the genetic tools that can be employed in the other models, analyzing the migration of zebrafish PGCs has the advantage of being able to follow the process at a very high spatial and temporal resolution in a live vertebrate embryo. We refer to and compare those findings with relevant data gathered in other models for single-cell migration.

This review presents the fundamental mechanisms that facilitate germ cell migration and arrival at their target rather than the full spectrum of the molecules involved. For a comprehensive review of all currently known activities and processes that influence the migration of germ cells in different organisms, see (Gross-Thebing et al., 2017; Oosterhuis & Looijenga, 2019; Richardson & Lehmann, 2010).

2. Motility of PGCs - principles and mechanisms

The movement of PGCs toward the region where the gonad develops takes place via different processes and distinct steps (Barton, Roa-de la Cruz, Lehmann, & Lin, 2024). For example, in *Drosophila*, certain migration steps are characterized by the attachment of the cells to somatic tissues that perform morphogenetic movements and carry the PGCs along; Most of the distance that *Drosophila* PGCs travel and their internalization deeper in the embryo are powered by gastrulation movement of the posterior midgut primordium (Warrior, 1994). Similarly, in the early zebrafish embryo, dorsal tissue convergence brings the PGCs toward medial positions (Weidinger, Wolke, Köprunner, Klinger, & Raz, 1999). Since, in such cases the surrounding tissues dictate the direction and extent of movement, the process is often referred to as "passive migration". In contrast, "active migration" is defined as the movement of the cells relative to surrounding cells and structures. This process typically depends on PGCs forming protrusions and engaging in dynamic, more transient interactions with other cells. In addition, one has to distinguish processes that are directly relevant to the migration of the cells from processes such as defects in cell fate and survival (Gross-Thebing et al., 2017; Weidinger et al., 2002). Last, active migration can be controlled by cues provided by other cells (e.g., Doitsidou et al., 2002) or by intrinsic cellular mechanisms (e.g., Hartwig et al., 2014).

Active migration is often preceded by a reduced expression level of adhesion molecules and the apolar formation of protrusions (Anderson, Copeland, Schöler, Heasman, & Wylie, 2000; Baronsky et al., 2016; Blaser et al., 2005; Goudarzi et al., 2012; Hartwig et al., 2014; Kunwar, Starz-Gaiano, Bainton, Heberlein, & Lehmann, 2003). The apolar formation of protrusions that precedes motility also coincides with increased actin polymerization, myosin contractility, and increased hydrostatic pressure (Goudarzi et al., 2012). When coordinated and directed, these activities can lead cells to persistently migrate in a random direction or a direction controlled by guidance cues (see "Directing PGC migration").

After acquiring motility, cellular protrusions form and serve various roles in migration and are generally considered required for motility. Interestingly, germ cells can produce two types of protrusions, namely pseudopods and blebs. These protrusion types differ in their kinetics and morphology, which is dictated by the underlying mechanisms that drive their formation. Pseudopods are powered by active actin polymerization at the leading edge, incrementally pushing the plasma membrane forward. Actin polymerization itself can be regulated by different molecules, such as the small GTPase Rac1, WASP/WAVE and the Arp2/3 complex (Devreotes & Horwitz, 2015). Bleb formation on the other hand relies on actomyosin contraction (García-Arcos, Jha, Waterman, & Piel, 2024; Schick & Raz, 2022). This contraction induces hydrostatic pressure, which leads to the delamination of the plasma membrane from the underlying actin cortex. Accordingly, bleb formation depends on the activity of proteins such as the small GTPase RhoA, which regulates actomyosin function through the kinase ROCK, which in turn induces phosphorylation of the regulatory light chain of non-muscle myosin 2 (Ridley, 2001). While hydrostatic pressure is the driving force, bleb formation is also influenced by other factors, such as the integrity of the actin cortex (Goudarzi et al., 2012; Paluch, Piel, Prost, Bornens, & Sykes, 2005; Sedzinski et al., 2011) and the level of the attachment between the plasma membrane and the actin cortex (Charras, Hu, Coughlin, & Mitchison, 2006; Olguin-Olguin et al., 2021).

Zebrafish PGCs mainly display blebs during active migration phases, although they can form pseudopods under certain conditions (Blaser et al., 2006; Truszkowski et al., 2023). Isolated *Xenopus, Drosophila,* and chick PGCs form bleb-like protrusions as well (Lin, Luo, & Lehmann, 2022;

Morita et al., 2023; Tarbashevich, Dzementsei, & Pieler, 2011; Terayama et al., 2013). However, in vivo, *Drosophila* and mouse PGCs appear predominantly spherical when migrating, showing protrusions only occasionally (Anderson et al., 2000; Seifert & Lehmann, 2012). Thus, the importance of protrusions for PGC migration in vivo has so far only been shown in zebrafish, where inhibiting the activity of RhoA abrogates both bleb formation and cell motility (Blaser et al., 2006).

In addi, reducing the activity of ano Rac1 activity, which consequently decreases actin polymerization, renders both zebrafish and mouse PGCs immotile (Jarysta et al., 2022; Kardash et al., 2010). In contrast, while the expression of dominant negative (DN)-RhoA also induces migration defects in *Drosophila* PGCs, the expression of a constitutively active form of Rac1, but not of DN-Rac1 impairs migration (Renault, Kunwar, & Lehmann, 2010).

As Rac1 and RhoA are dominant factors that control actin polymerization and contractility in germ cells, their activity is enriched at specific regions of the cell, namely the front and rear aspects respe (Kardash et al., 2010). For example, activated RhoA has been detected at the leading edge of cells, where bleb formation occurs, as well as at the cell rear where it could be required for retraction as shown for neutrophils (Blaser et al., 2006; Hind, Vincent, & Huttenlocher, 2016).

Unlike RhoA, Rac1 activation is detected only at the front of migrating zebrafish PGCs, an area also displaying increased actin polymerization (Kardash et al., 2010). Interestingly, it has been reported that an increase in actin polymerization at the front and the initiation of actin retrograde flow are the first activities observed as zebrafish PGCs transition from an apolar immotile state to a polar migratory state (Fig. 1A) (Olguin-Olguin et al., 2021). Indeed, light-induced activation of Rac1 increases actin polymerization and actin retrograde flow at the induction site, which becomes the cell front (Olguin-Olguin et al., 2021). Retrograde flow of actin results from the polymerization of actin filaments that push against the front and the contraction of myosin that leads to flow of actin filaments backward (Fig. 1A) (Henson et al., 1999; Mitchison & Kirschner, 1988; Pollard, Blanchoin, & Mullins, 2000). Carried along this flow are molecules with an affinity for actin, and they accumulate at the back of cells. In the case of zebrafish PGCs, such molecules include the membrane-cortex linker molecule Ezrin and the membrane-endoplasmic reticulum connector extended synaptotagmin-like 2a (Esyt2a) (Olguin-Olguin et al., 2021). These linkers inhibit blebbing at the rear while allowing for protrusion

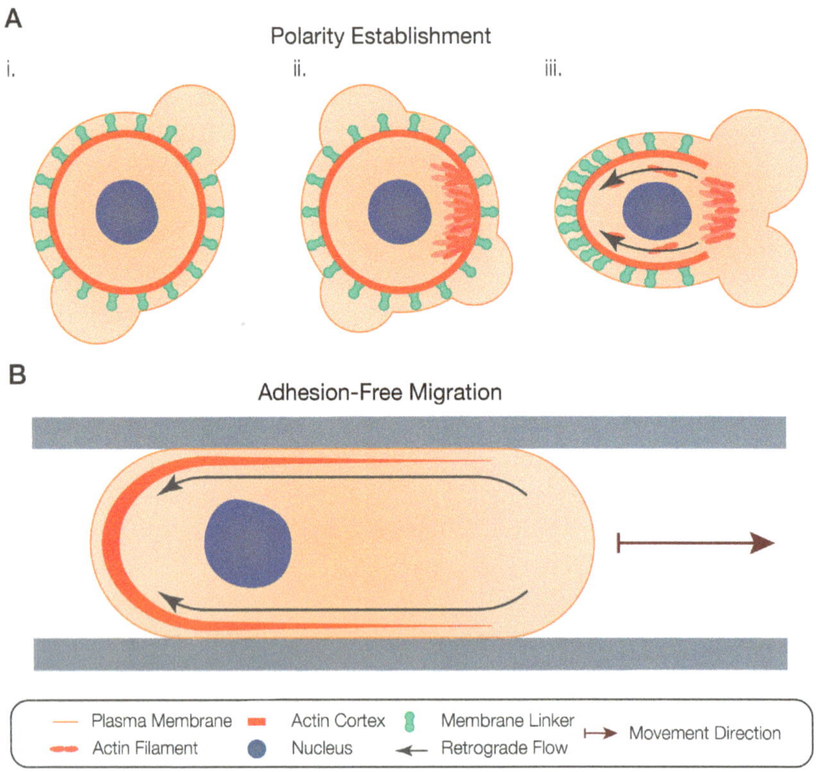

Fig. 1 Polarity establishment in migrating zebrafish PGCs. (A) (i) Before motility takes place, linker molecules are distributed homogenously on the membrane, and protrusions are formed all around the cell perimeter. (ii) Actin polymerization defines the front aspect of the cell. (iii) Retrograde flow of actin filaments results in accumulation of membrane linker proteins at the opposing aspect thereby defining the rear of the cell and biasing the formation of protrusions to the front. (B) Adhesion-free migration by retrograde actin flow. Combined action of actin polymerization at the front and actomyosin contraction induce retrograde flow of actin filaments towards the rear of the cell. Forces resulting from retrograde flow are transmitted to the substrate through nonspecific frictional interactions leading to forward translocation of the cell.

formation at the front. Indeed, overexpression of a hyperactive form of Ezrin reduces blebbing activity (Fig. 1A) (Olguin-Olguin et al., 2021). Further, in zebrafish PGCs, abrogating actomyosin contractility by the expression of DN-ROCK inhibits actin retrograde flow. Under such conditions, the front of cells still shows the characteristic actin polymerization, but the linker molecules Ezrin and Esyt2a are not enriched at the cell back (Olguin-Olguin et al., 2021).

Drosophila PGCs also exhibit actomyosin-dependent actin retrograde flow (Lin et al., 2022) as well as the associated polar accumulation of specific proteins at the rear of cells (Kunwar et al., 2008). An additional mechanism that affects the speed of actin retrograde flow involves the function of E-cadherin engagement; In zebrafish PGCs, the levels of engaged E-cadherin were shown to control the flow speed (Grimaldi et al., 2020).

In addition to its role in polarity, actin retrograde flow could also be the driving force for the actual movement of amoeboid cells, such as PGCs (Paluch, Aspalter, & Sixt, 2016). The retrograde flow-powered migration for amoeboid cells differs from that suggested for mesenchymal cells, in which the prevailing model for migration is the "molecular clutch" mechanism (Case & Waterman, 2015; Elosegui-Artola, Trepat, & Roca-Cusachs, 2018). This model proposes that cells utilize protrusions to expand the leading edge in the direction of migration and adhere to the extracellular matrix (ECM), followed by actomyosin contraction-generated pulling forces. The result of this cycle is the forward movement of the cell. While this model generally applies to mesenchymal migration, the role protrusions play in amoeboid migration is less clear, since these cells typically exhibit a reduced requirement for adhesion (Lämmermann & Sixt, 2009; Schick & Raz, 2022).

In the case of amoeboid cells, retrograde flow may contribute to friction through non-specific interactions with the substrate, which would propel the cell forward (Fig. 1B) (Hawkins et al., 2011; Ruprecht et al., 2015). The friction could arise from any cell surface molecule interacting directly or indirectly with actin filaments that undergo retrograde flow. Evidence for this mechanism was initially obtained in an in vitro setting of migrating amoeboid cells, where the friction patterns exerted on the substrate were consistent with those expected from non-specific friction-based migration in a non-adhesive environment (Bergert et al., 2015). Direct evidence for this mechanism is still lacking in PGCs. However, inhibiting the function of cell-surface adhesion molecules in zebrafish PGCs did not abrogate migration. Specifically, zebrafish PGCs in which integrin β1a and integrin β1b were knocked down arrived at the target, while showing migration kinetics similar to those in non-manipulated cells (Kardash et al., 2010). Similarly, impairing integrin function by introducing the RGD peptide into the zebrafish embryo has resulted in severe developmental defects, while PGCs still migrated to form clusters at the site where the gonad should develop (Kardash et al., 2010). In mice, however, PGCs lacking integrin β1 showed migration defects, albeit very mild, with successful arrival at the target (Anderson et al., 1999; Carver, He, & Zhu, 2021). The seemingly different roles cell-ECM adhesion plays in PGC migration could stem from the need for cells to cope with diverse migration paths and tissues they encounter in different organisms. However, it remains an open question whether defects such as these interfere with the ability of PGCs to migrate or whether they represent strong, non-specific, and pleiotropic changes that are not directly related to the mechanisms of migration processes under physiological conditions.

In addition to the effect of ECM and integrins on PGC migration, cell-cell interactions also play a role in migration processes, for example by connecting PGCs to their neighbors during passive migration (Chihara & Nance, 2012; DeGennaro et al., 2011), as well as by controlling the onset of active PGC migration (Molyneaux, Stallock, Schaible, & Wylie, 2001; Seifert & Lehmann, 2012). The cell-cell adhesion molecule E-cadherin is involved in migration processes in an additional way: As mentioned above, knocking down E-cadherin activity in zebrafish germ cells was found not to affect active migration per se but to reduce directional persistence. Specifically, reducing E-cadherin activity allowed actin flow to proceed faster, such that the front, as defined by the distribution of actin, extends further toward the cell back. Under these conditions, the formation of protrusions is less focused, such that the migration path is less straight (Grimaldi et al., 2020).

Establishing the polarity axis and the robust mechanisms that maintain it allow cells to migrate within embryonic tissues that exhibit different properties, such as the level of stiffness (Krieg et al., 2008). While PGCs can migrate within various tissues, their positioning is directed by the distribution of the guidance cue Cxcl12 (Gross-Thebing et al., 2020). Interestingly, PGCs show different migration characteristics when located within ectodermal or endodermal tissues (Truszkowski et al., 2023). Whereas the migration speed and track straightness remain unchanged, the frequency of specific protrusion types correlates with the properties of the surrounding tissue. In mesodermal tissues, PGCs form blebs more frequently, while pseudopods are more common in PGCs that migrate within the ectoderm. Adhesion, confinement, and substrate stiffness could be the tissue properties that influence the protrusion types formed by amoeboid cells (Schick & Raz, 2022).

3. Tissue barriers in PGC migration

The migration of PGCs is directed not only by chemoattractants but also by tissues that the cells stay out of. Such tissues can act as physical barriers or emit repulsive signals that direct the migration away from them. While such barriers affect the migration upon physical interactions, repulsive signals can also function at a distance.

Physical barriers can limit the entry of cells into specific areas within the embryo by forming structures that cannot be crossed due to their physical properties. For example, cells cannot easily pass into tissues in which the cells have strong stable connections or when pore sizes are inadequate (Friedl, Wolf, & Lammerding, 2011; Garcia-Arcos et al., 2019; McGregor, Hsia, & Lammerding, 2016). In *Drosophila,* PGCs have been found to cross physical barriers, such as their migration through the midgut epithelium (Fig. 2A). Interestingly, this migration step depends on the remodeling of the endodermal epithelium by itself, which becomes permissive for migration of the PGCs in between gaps that form among cells. Accordingly, when endoderm differentiation is impaired, PGCs cannot pass through it on their way to their migration target (Jaglarz & Howard, 1995; Seifert & Lehmann, 2012). Notably, the alterations in the properties of the endoderm are not achieved through active action by the PGCs but are a result of a developmental process independent of interactions between the PGCs and the tissue to be transversed. Thus, this event is different from that in which leukocytes trigger the remodeling of connections among cells within the tissue they migrate through or even exhibit migration through the cells themselves (Nourshargh & Alon, 2014).

In addition to being structures the cells have to pass, physical barriers also regulate cell positioning. For instance, zebrafish PGCs that make contact with the gut display a reversal in their polarity, which leads to migration away from the structure (Paksa et al., 2016). Thus, the developing gut represents a physical barrier that separates the two PGC clusters and disruption of gut development results in their fusion into one cluster (Fig. 2B) (Paksa et al., 2016). The basis for this behavior, is a reduction in the level of polymerized actin at the leading edge upon contact, followed by the generation of an actin-rich new front (Fig. 2C). Similarly, PGCs interacting with the notochord, show a decrease in the level of polymerized actin at the contact site and subsequently migrate along or away from it (Gross-Thebing et al., 2020). Here, the angle of the new front depends on the angle at which the cell contacts the notochord. Encountering an obstacle could lead to decreased actin polymerization at the contact site, while continuous actin polymerization at parts of the leading edge that do not contact the obstacle forms a front in a new direction (Fig. 2C). In the case of the notochord the sensing of the barrier was suggested to be a

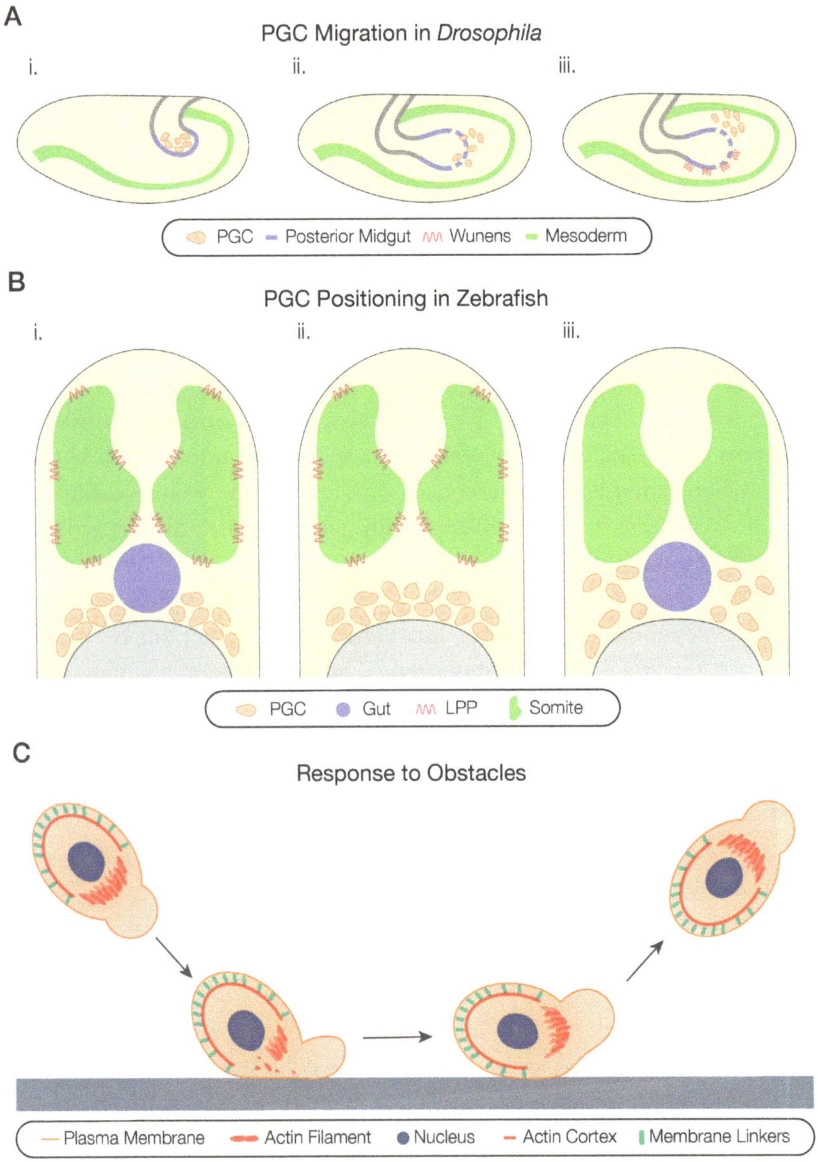

Fig. 2 (A) (i) *Drosophila* PGCs are passively translocated during gastrulation together with the involuting posterior midgut primordium. (ii) Formation of gaps within the midgut of the endodermal epithelium allows for the migration of PGCs through. (iii) The activity of Wunen proteins at the ventral part of the midgut directs the movement of PGCs toward the mesoderm. B (i) In wildtype zebrafish embryos PGCs form two separate clusters on either side of the gut. The cells are restricted to ventral regions by the activity of LPPs. (ii) In the absence of gut tissue, the PGC clusters fuse at the midline of the embryo. (iii) Lack of LPP function leads to positioning of PGCs at more

response to the high stiffness of the ECM around the structure (Gross-Thebing et al., 2020). Along these lines, in one study that monitored the behavior of PGCs upon interacting with biologically inert hydrogels, it was demonstrated that PGCs react to purely physical cues (Truszkowski et al., 2023). Furthermore, this study showed that not only could the cells sense and respond to physical cues, but they also exhibited varying responses to different physical properties of the structure. Specifically, PGCs encountering stiffer hydrogels showed a higher rate of polarity loss than those encountering softer gels (Truszkowski et al., 2023). In those cases, the basis for the change in migration direction may involve signaling cascades, especially in the case of sensing the obstacle's degree of stiffness (Trappmann & Chen, 2013). Thus, the responses of the PGCs to interactions with the gut or the notochord could involve detecting stiffness or recognizing key biochemical properties of the barrier, such as E-cadherin activity levels. For example, zebrafish PGCs that encounter regions devoid of E-cadherin respond by turning, which was suggested to be caused by asymmetrical engagement between E-cadherin molecules of the PGCs and their neighbors (Grimaldi et al., 2020). This situation could lead to more rapid actin retrograde flow at sites where E-cadherin engagement is lower, thereby effectively decreasing actin concentration at this aspect of the cell. Consequently, the side of the cell where actin concentration is higher would then become the site at which a new leading edge forms, inducing movement away from the E-cadherin-depleted areas.

4. Directing PGC migration

In the context of development, cell motility is, in most cases, directed toward specific destinations where the cells contribute to formation of tissues and organs (e.g., Szabó & Mayor, 2018) like the gonad in the case of PGCs. It is important to note, that failure to reach the migration target can result from impaired motility or from defects in perceiving and responding to guidance cues. For example, the Receptor Tyrosine Kinase c-KIT and its ligand Steel were previously believed to provide directional guidance to mouse PGCs, as a loss of either was found to result in decreased efficiency of

dorsal locations. (C) Upon interaction with physical barriers, actin polymerization is reduced at the contact site. Conversely, actin polymerization at the aspect of the leading edge that does not come in contact with the barrier, defines a new front, resulting in migration away from the obstacle.

germ cells to arrive at the genital ridges (Buehr, McLaren, Bartley, & Darling, 1993; Godin et al., 1991). However, it has since been shown that c-KIT/Steel signaling promotes motility rather than provide guidance. In embryos deficient in Steel, germ cells migrate directionally but at significantly reduced speeds (Gu, Runyan, Shoemaker, Surani, & Wylie, 2009; Gu, Runyan, Shoemaker, Surani, & Wylie, 2011). To distinguish between the roles of guidance cues and motility promoters, it is crucial to measure dynamic parameters such as migration speed and directionality, rather than solely assessing the efficiency of target arrival. Below, we present findings relevant to directing the migration of germ cells toward their targets.

During embryonic development, various cell types migrate to distinct locations simultaneously, which calls for mechanisms facilitating specific responses to specific signals. In the case of secreted guidance cues, the range by which the ligand is distributed should be limited (Boldajipour et al., 2008). An additional mechanism that contributes to chemokine signaling specificity relies on the expression of receptors that can bind and transmit signals encoded by specific ligands, but not others (Malhotra, Shin, Solnica-Krezel, & Raz, 2018).

While notable differences exist among organisms regarding how and where germ cells are specified, the primary mechanisms guiding the migration of PGCs are similar in common model organisms such as *Drosophila*, zebrafish, and mice. These mechanisms predominantly involve responses to soluble guidance cues and reactions toward physical barriers (Barton, LeBlanc, & Lehmann, 2016; Richardson & Lehmann, 2010).

In particular, chemokine receptors of the CXCR group and their ligands (specifically, Cxcr4 and Cxcl12) have been shown to direct vertebrate germ cells toward their target in zebrafish, chick, *Xenopus*, and mouse among others (Fig. 3) (Ara et al., 2003; Doitsidou et al., 2002; Molyneaux et al., 2003; Stebler et al., 2004; Takeuchi, Tanigawa, Minamide, Ikenishi, & Komiya, 2010). Chemokines are small, secreted proteins that activate their cognate receptor, which signals by stimulating GDP/GTP exchange of coupled G proteins and dissociation of the α and $\beta\gamma$ subunits. This then activate downstream signaling pathways that direct cell polarization and migration toward the source of the ligand (Bachelerie et al., 2014; Hughes & Nibbs, 2018; Zlotnik & Yoshie, 2012).

Guiding zebrafish germ cells is an especially challenging task, since the cells are specified at four different locations within the embryo. Yet, within 24 hours, all the cells cluster at the location where the gonad forms (Weidinger et al., 1999, 2002; Yoon, Kawakami, & Hopkins, 1997). In the zebrafish model it was first demonstrated that Cxcl12 and Cxcr4 are critical for guiding the PGCs to their target. Indeed, in embryos lacking the receptor or its ligand, the cells arrive at ectopic locations (Doitsidou et al., 2002; Knaut, Werz, Geisler, & Nüsslein-Volhard, 2003). Consistent with the proposed function of Cxcl12 as the guidance cue, it was shown to be expressed in somatic cells along the migration path of the cells, while Cxcr4 is expressed on the membrane of the migrating cells themselves (Doitsidou et al., 2002). Significantly, ectopic expression of Cxcl12 was shown to be sufficient to guide germ cells toward these locations, further demonstrating the role of the protein as an instructive, attractive cue that is both required and sufficient to direct PGC migration (Doitsidou et al., 2002). Thus, studies in zebrafish have clearly shown that Cxcl12 signaling through Cxcr4 acts specifically as a guidance cue for motile cells. Further work has shown that it functions in determining the location where the cell front is generated (Fig. 3) (Olguin-Olguin et al., 2021).

Following the findings in zebrafish, the involvement of the Cxcr4/Cxcl12 module was examined in other vertebrates. PGCs exhibit migration defects in mouse in which Cxcr4 is mutated (Ara et al., 2003; Molyneaux et al., 2003). Similar to observations in zebrafish, in mouse, the loss of a ligand or receptor results in PGCs failing to colonize the genital ridges, and ectopic expression of the attractant is sufficient to mislocalize germ cells (Ara et al., 2003; Molyneaux et al., 2003). Indeed, mouse PGCs actively migrate through the hindgut toward the genital ridges, where the chemokine is highly expressed (Anderson et al., 1999). In chick embryos, PGCs migrate toward regions where the ligand is expressed and toward domains where it is experimentally introduced (Takeuchi et al., 2010). Similarly, in *Xenopus* knocking down the function of the receptor resulted in impaired migration of PGCs.

To date, chemokines and their receptors have not been found in invertebrates. Instead, in *Drosophila*, studies have identified a range of molecules whose function is required for the proper arrival of PGCs to the gonad. Among those molecules, the GPCR Trapped in endoderm 1 (Tre1) was found to be required within the PGCs themselves for successful transepithelial migration through the posterior midgut; as the name implies, *Drosophila* PGCs deficient in Tre1 fail to transverse the midgut and remain trapped within the epithelium (Kunwar et al., 2003, 2008). The role of Tre1 in controlling directional migration is thought to involve binding one or more guidance cues (Coffman et al., 2002; Kim, Hanlon, Vohra, Devreotes, & Andrew, 2021; LeBlanc & Lehmann, 2017). Furthermore,

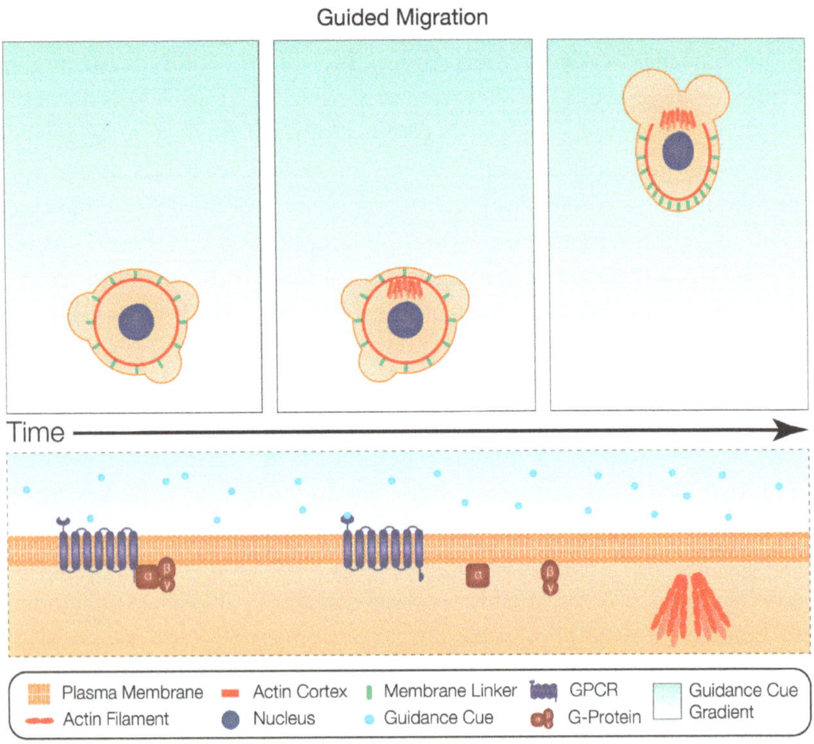

Fig. 3 Response of cells to guidance cues. Cells can polarize and migrate towards a source of an attractant. Higher activation levels of G-protein coupled receptors (GPCRs) at one aspect of the cell induces intracellular signaling cascades that enhance polymerization of actin filaments in the direction of the chemokine gradient. Actin polymerization at the cell front, actin retrograde flow, and subsequent linker accumulation at the rear allow cells to migrate up the gradient.

studies have revealed that the Hedgehog (Hh) signaling pathway and its upstream regulator Hmgcr, the *Drosophila* homolog of HMG-CoA, provide directional guidance to PGCs. Like Cxcl12 in vertebrates, ectopic expression of Hmgcr or Hh was sufficient to attract PGCs in *Drosophila* (Deshpande et al., 2023; Van Doren, Broihier, Moore, & Lehmann, 1998). Recent findings connect these fundamental discoveries functionally by suggesting that Hh signaling increases levels of Tre1 at the plasma membrane through the seven-pass membrane protein Smoothened (Smo) (Kim et al., 2021). In vertebrates, inhibition of HMG-CoA reductase causes comparable migration defects and mislocalization of germ cells; however, its expression is insufficient to provide any directional information (Thorpe, Doitsidou, Ho, Raz, & Farber, 2004). Still, the common relevance of HMG-CoA and Hh signaling and the key function of GPCRs implies high conservation of PGC guidance strategies across species.

Like other chemokines, the distribution of Cxcl12 is regulated not only by its RNA expression pattern but also by interaction with the ECM (e.g., Murphy et al., 2007) and, importantly, also by sequestration of the ligand by a decoy receptor, Cxcr7 (Boldajipour et al., 2008). Like the phenotype observed in embryos lacking Cxcr4, in embryos knocked down for Cxcr7 the germ cells fail to reach the gonad. However, comparing the behavior of cells lacking Cxcr4 with those knocked down for Cxcr7 revealed a significant difference. While PGCs deficient for Cxcr4 were motile, those with reduced levels of Cxcr7 remained at one spot while extending protrusions in all directions (Boldajipour et al., 2008). The basis for this phenotype was found to be the elevated levels of Cxcl12 in the extracellular space because it could not be internalized by the receptor (Boldajipour et al., 2008). Thus, the function of Cxcr7 allows changes in the highly dynamic RNA expression pattern of Cxcl12 to be mirrored by corresponding changes in the protein distribution.

In addition to chemical cues that function as attractants, signaling cascades were also shown to direct migration away or repel cells from specific locations. For example, semaphorins and ephrins constitute repulsive signals that guide the neuronal growth cone (Hinck, 2004). The expression of Wunens (lipid phosphate phosphatases (LPPs)) at the ventral part of the midgut are thought to direct the migration of PGCs towards the mesoderm (Fig. 2A) (Richardson & Lehmann, 2010; Starz-Gaiano, Cho, Forbes, & Lehmann, 2001). *Drosophila* PGCs avoid entering regions that express Wunen and Wunen2 in wild-type embryos as well as domains in which these enzymes are induced experimentally (Starz-Gaiano et al., 2001; Zhang, Zhang, Purcell, Cheng, & Howard, 1997). Consistent with the idea that Wunens guide PGCs, genetic mutations in their genes result in PGC migration defects (Starz-Gaiano et al., 2001; Zhang et al., 1997). Like other LPPs, Wunens are thought to act through the dephosphorylation of extracellular phospholipids, which in this case abrogates protein function (Burnett & Howard, 2003; Pyne, Kong, & Darroch, 2004; Starz-Gaiano et al., 2001). Analogous to Wunens in *Drosophila,* LPPs in zebrafish regulate the position of PGCs. When LPPs are knocked down or knocked out, zebrafish PGCs localize to positions closer to somites, an LPP-expressing tissue (Fig. 2B). In the same vein, ectopic expression of LPPs in specific embryonic regions causes migrating PGCs to avoid them (Paksa et al., 2016).

Importantly, however, it remains unclear whether Wunens generate a repulsive signal per se or whether they degrade an attractant, thereby generating an inverse attractant gradient and resulting in migration away from this area (Hanyu-Nakamura, Kobayashi, & Nakamura, 2004; Renault & Lehmann, 2006; Renault, Sigal, Morris, & Lehmann, 2004).

5. Increasing migration precision

Establishing and maintaining front-rear polarity is required for the persistent movement of cells in specific directions. As described above, the distinction between the front and rear aspects of the cell is maintained by antagonistic biophysical and biochemical activities operating on either side of the cell. For example, in zebrafish, inhibition of bleb formation at the cell rear by linker molecules, such as Ezrin and Esyt2a, that are transported away from the front, coupled with enhanced levels of actin polymerization at the cell leading edge, give rise to the definition of a very robust axis of migration (Olguin-Olguin et al., 2021). An even more robust front-rear polarity is observed in migrating bacteria. In this case, the force propelling the cell forward is oriented strictly in the direction of the long axis of the bacterial cells that maint fixed morphology (Wadhwa & Berg, 2022). While the polarized distribution of cellular structures and molecules is essential for persistent directional migration, these elements should allow flexibility, thereby permitting corrections to the migration path. For example, adjustments to the direction of migration are needed for moving targets or when the original migration track is not correctly oriented, as is likely when guidance cues form shallow gradients. Changes in the direction of cell migration are also required for efficient scouting of tissues, for example, when scanning large areas for the presence of pathogens or tissue damage and to colonize new territories during embryogenesis (Petrie, Doyle, & Yamada, 2009).

An interesting phenomenon associated with changes in migration direction, is that the robust and persistent polarity is periodically abolished. The effect of such a polarity loss events is a stop in migration, often followed by a change in the direction of migration. This behavior, termed "run and tumble," has been described in cells ranging from small bacteria to large eukaryotic cells such as germ cells and leukocytes, as they respond to chemical and mechanical directional cues (Berg & Brown, 1972; Georgantzoglou, Poplimont, Walker, Lämmermann, & Sarris, 2022; Reichman-Fried, Minina, & Raz, 2004; Zhang et al., 2023).

In *E. coli*, where run and tumble behavior was first discovered, the cellular mechanism underlying the transition between the two behavior modes is the direction of flagella rotation and the degree of bundling of filaments that propel the cell forward (Wadhwa & Berg, 2022). In bacteria swimming up gradients of attractants, the duration of filament bundling is prolonged, resulting in extended run phases. Conversely, unbundling, which causes tumbling, occurs more often for bacteria that move down a gradient of an attractant. Since the run phases are longer when swimming up the gradient and the direction of exit from the tumbling is random, bacteria perform "biased random walks" that bring the cells closer to the attractant source. This strategy allows cells with a size of 1–2 μm to detect and respond to the orientation of the chemoattractant gradient by acting as sensors that compare the level of nutrients in time while moving. In contrast, migrating eukaryotic cells such as PGCs are large enough to detect differences in the concentration of the attractant across their length.

In the case of eukaryotic cells, erasing intracellular polarity during tumbling phases is thought to render the cells naïve. As a result, changes to the migration path are not being opposed by the previously established front-rear axis. Accordingly, in the presence of a gradient of the attractant, the direction in which the cells exit tumbling phases is biased by the distribution of the cue (Fig. 4A). Thus, while in bacteria the direction of the guidance cue gradient is detected during run phases, zebrafish germ cells first decode the Cxcl12 gradient toward the end of tumbling phases and initiate the run phase in the direction of the chemokine source.

Similar to the inherent motility of zebrafish germ cells, "run and tumble" is an autonomous behavior that is also observed when the germ cells are positioned within different tissue environments and in the absence of the guidance cue or its receptor (Reichman-Fried et al., 2004). However, the run and tumble behavior is influenced by chemokine signaling in a way that increases the accuracy of the migration process and arrival at the developing gonad region (Minina, Reichman-Fried, & Raz, 2007; Reichman-Fried et al., 2004). The involvement of chemokine signaling in regulating run and tumble behavior was revealed when PGC migration was investigated in embryos in which Cxcl12 or its receptor Cxcr4 were knocked down (Reichman-Fried et al., 2004). In such embryos, the duration of run phases was reduced. Thus, after directing migration at the exit from tumble toward the high point of the Cxcl12 gradient, chemokine signaling was able to control the duration of the active migration phases, thereby increasing the overall speed by which PGCs approached their targets.

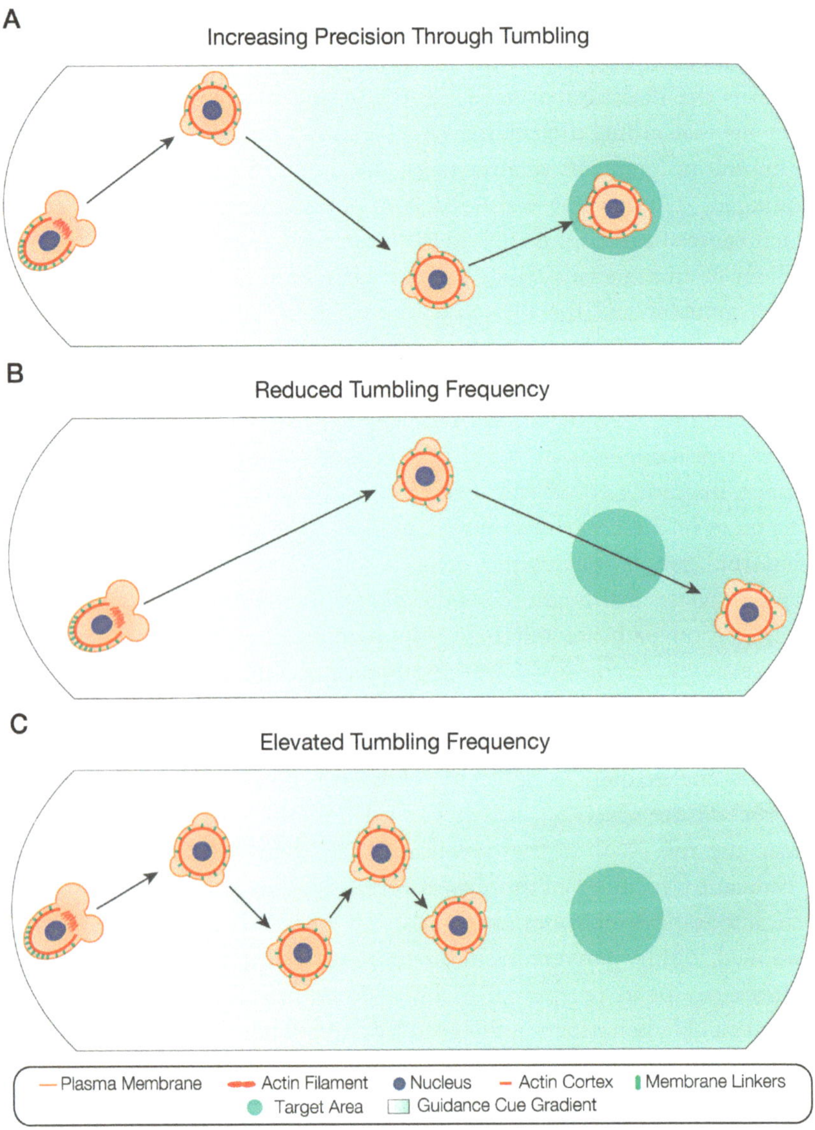

Fig. 4 (A) Periodic loss of polarity by tumbling renders the cells naïve, allowing them to correct their migration path and ensure proper arrival at the target area. The frequency of tumbling events can affect the proper arrival and stop at the target area. (B) If tumbling is not frequent enough, cells might overshoot the target area. (C) Too frequent tumbling might lead to reduced displacement that results in inability to reach the target area in time.

While a longer run duration is advantageous as the cells directionally migrate up the chemokine gradient, suppressing the run phase length is required when cells come closer to their target to avoid migrating beyond their destination (Fig. 4B). Indeed, studies have shown that the duration of run phases was dramatically reduced as the cells came closer to positions containing higher levels of the expressed chemokine (Minina et al., 2007). The mechanism by which PGCs sense their arrival at the vicinity of the target involves the degree of chemokine receptor internalization, which is elevated in regions of high ligand levels. Thus, germ cells that cannot properly internalize Cxcr4, often pass by their correct target and arrive at ectopic positions in the embryo (Fig. 4B) (Minina et al., 2007). The reduction in the length of run phases is an important final step of the migration as the cells arrive at the target. However, controlling the duration of run and tumble phases is important also at other stages of the migration process. For example, too frequent tumble phases would lead to a reduction in cell displacement that would not allow the cells to arrive at their target in time (Fig. 4C).

In zebrafish PGCs, the control by the guidance cue over the run phase length resembles some features of bacterial chemotaxis, particularly the mechanism they use to reach the target and the increased run phase duration induced by the ligand. While the molecular mechanisms controlling the transitions between run and tumble phases are well characterized in bacteria, they are poorly understood in germ cells. In particular, the mechanisms that facilitate the abolishment of the front-rear polarity axis, entry into tumble phases, or stopping at the target are unknown. Previous studies have shown that at the exit from tumble phases, the establishment of a cell front is the event that initiates the polarization and dictates where the back will form (Olguin-Olguin et al., 2021). Conversely, when the cell rear is defined first, such as following cell division, this region becomes the instructive entity, and the cell leading edge forms on the opposite side. Uncovering the molecular cascade leading to the elimination of polarity and entry into tumbles would be an important topic for future investigation. Similarly, an interesting line of research might focus on determining the molecular cascades that result in periodic gain and loss of PGC polarity and how external factors such as ligand distribution and level affect it.

6. Perspectives

Germ cell migration is an excellent in vivo model for analyzing cell motility, directed migration, and interaction with physical structures in embryonic tissue. However, the characterization of the process concerning dynamic changes in cell morphology, subcellular organization of molecules, activities, and position of organelles during migration are less detailed than findings obtained in in vitro models for cell migration. This is due to the additional challenges arising from in vivo setups where migration is studied within tissues, such that the chemical and physical conditions are not uniform and are more difficult to control. In the case of in vivo studies, tissue properties are more heterogeneous, the cells are located deep within tissues that themselves undergo morphogenetic movements. These difficulties strongly affect researchers' ability to acquire high spatiotemporally resolved data from the relatively small number of cells per embryo (e.g., about 15 in the zebrafish embryo at the relevant stages). Yet, technological advances in microscopy could help by increasing tissue penetration and allowing for faster imaging; further, the field would benefit considerably from the development of better analytical tools. Such tools could improve image quality and enable the definition of shapes and position of cellular structures by employing image segmentation, which is essential when studying germ cell migration in the three-dimensional context of the embryo. Critically, computational approaches that use processed image data could assist in identifying relationships among cellular components and cell-shape features, offering a holistic understanding of the migration process. Obtaining and analyzing such information promises to provide an unbiased view of germ cell migration, which has, thus far, been characterized by very few parameters that were measured employing partially subjective criteria (Meyen et al., 2015; Truszkowski et al., 2023).

Following these comprehensive approaches in embryos in which cells are located within different embryonic environments, subjected to varying levels of soluble cues, interacting with physical structures, etc., will provide further insights into the mechanisms controlling cell migration. These advances will likely influence research on other cell types by offering tools and establishing paradigms that may be universally relevant.

An additional topic that needs to be better understood relates to the ability of PGCs to preserve their fate during migration. Germ cells are specified during the early stages of embryonic development when extensive differentiation processes take place. Nevertheless, as shown most dramatically in zebrafish (Gross-Thebing et al., 2020; Weidinger et al., 2002), germ cells maintain their fate despite interacting with essentially all the cell types of the early embryo. Investigations into the molecular mechanisms contributing to this property of germ cells and the consequences of their failure are important future tasks.

Last, the study of germ cell migration in vivo has provided a large body of information concerning the topics presented in this article. Thus, it would be exciting to examine the extent to which the mechanisms identified in the main experimental models are conserved with those that function in "non-model systems" (Extavour & Akam, 2003).

Acknowledgments

We thank Celeste Brennecka for editing and Michal Reichman-Fried for critical comments on the manuscript. The work in the lab is supported by funding from the University of Münster, the Max Planck Institute for Molecular Biomedicine and the German Research Foundation (DFG) grant RA 863/14-1 and SFB 1348 project B06.

References

Anderson, R., Copeland, T. K., Schöler, H., Heasman, J., & Wylie, C. (2000). The onset of germ cell migration in the mouse embryo. *Mechanisms of Development, 91*, 61–68. https://doi.org/10.1016/S0925-4773(99)00271-3.

Anderson, R., Fässler, R., Georges-Labouesse, E., Hynes, R. O., Bader, B. L., Kreidberg, J. A., ... Wylie, C. (1999). Mouse primordial germ cells lacking β1 integrins enter the germline but fail to migrate normally to the gonads. *Development (Cambridge, England), 126*, 1655–1664. https://doi.org/10.1242/dev.126.8.1655.

Ara, T., Nakamura, Y., Egawa, T., Sugiyama, T., Abe, K., Kishimoto, T., ... Nagasawa, T. (2003). Impaired colonization of the gonads by primordial germ cells in mice lacking a chemokine, stromal cell-derived factor-1 (SDF-1). *Proceedings of the National Academy of Sciences, 100*, 5319–5323. https://doi.org/10.1073/pnas.0730719100.

Bachelerie, F., Ben-Baruch, A., Burkhardt, A. M., Combadiere, C., Farber, J. M., Graham, G. J., ... Zlotnik, A. (2014). International union of pharmacology. LXXXIX. Update on the extended family of chemokine receptors and introducing a new nomenclature for atypical chemokine receptors. *Pharmacological Reviews, 66*, 1–79. https://doi.org/10.1124/pr.113.007724.

Baronsky, T., Dzementsei, A., Oelkers, M., Melchert, J., Pieler, T., & Janshoff, A. (2016). Reduction in E-cadherin expression fosters migration of Xenopus laevis primordial germ cells. *Integrative Biology, 8*, 349–358. https://doi.org/10.1039/c5ib00291e.

Barton, L. J., LeBlanc, M. G., & Lehmann, R. (2016). Finding their way: Themes in germ cell migration. *Current Opinion in Cell Biology, Cell dynamics, 42*, 128–137. https://doi.org/10.1016/j.ceb.2016.07.007.

Barton, L. J., Roa-de la Cruz, L., Lehmann, R., & Lin, B. (2024). The journey of a generation: Advances and promises in the study of primordial germ cell migration. *Development (Cambridge, England), 151*, dev201102. https://doi.org/10.1242/dev.201102.

Berg, H. C., & Brown, D. A. (1972). Chemotaxis in Escherichia coli analysed by three-dimensional tracking. *Nature, 239*, 500–504. https://doi.org/10.1038/239500a0.

Bergert, M., Erzberger, A., Desai, R. A., Aspalter, I. M., Oates, A. C., Charras, G., ... Paluch, E. K. (2015). Force transmission during adhesion-independent migration. *Nature Cell Biology, 17*, 524–529. https://doi.org/10.1038/ncb3134.

Blaser, H., Eisenbeiss, S., Neumann, M., Reichman-Fried, M., Thisse, B., Thisse, C., & Raz, E. (2005). Transition from non-motile behaviour to directed migration during early PGC development in zebrafish. *Journal of Cell Science, 118*, 4027–4038. https://doi.org/10.1242/jcs.02522.

Blaser, H., Reichman-Fried, M., Castanon, I., Dumstrei, K., Marlow, F. L., Kawakami, K., ... Raz, E. (2006). Migration of zebrafish primordial germ cells: A role for myosin contraction and cytoplasmic flow. *Developmental Cell, 11*, 613–627. https://doi.org/10.1016/j.devcel.2006.09.023.

Boldajipour, B., Mahabaleshwar, H., Kardash, E., Reichman-Fried, M., Blaser, H., Minina, S., ... Raz, E. (2008). Control of chemokine-guided cell migration by ligand sequestration. *Cell, 132*, 463–473. https://doi.org/10.1016/j.cell.2007.12.034.

Buehr, M., McLaren, A., Bartley, A., & Darling, S. (1993). Proliferation and migration of primordial germ cells in W/W mouse embryos. *Developmental Dynamics, 198*, 182–189. https://doi.org/10.1002/aja.1001980304.

Burnett, C., & Howard, K. (2003). Fly and mammalian lipid phosphate phosphatase isoforms differ in activity both in vitro and in vivo. *EMBO Reports, 4*, 793–799. https://doi.org/10.1038/sj.embor.embor900.

Carver, J. J., He, Y., & Zhu, Y. (2021). Delay in primordial germ cell migration in adamts9 knockout zebrafish. *Scientific Reports, 11*, 8545. https://doi.org/10.1038/s41598-021-88024-x.

Case, L. B., & Waterman, C. M. (2015). Integration of actin dynamics and cell adhesion by a three-dimensional, mechanosensitive molecular clutch. *Nature Cell Biology, 17*, 955–963. https://doi.org/10.1038/ncb3191.

Charras, G. T., Hu, C.-K., Coughlin, M., & Mitchison, T. J. (2006). Reassembly of contractile actin cortex in cell blebs. *Journal of Cell Biology, 175*, 477–490. https://doi.org/10.1083/jcb.200602085.

Chihara, D., & Nance, J. (2012). An E-cadherin-mediated hitchhiking mechanism for C. elegans germ cell internalization during gastrulation. *Development (Cambridge, England), 139*, 2547–2556. https://doi.org/10.1242/dev.079863.

Coffman, C. R., Strohm, R. C., Oakley, F. D., Yamada, Y., Przychodzin, D., & Boswell, R. E. (2002). Identification of X-linked genes required for migration and programmed cell death of Drosophila melanogaster germ cells. *Genetics, 162*, 273–284. https://doi.org/10.1093/genetics/162.1.273.

DeGennaro, M., Hurd, T. R., Siekhaus, D. E., Biteau, B., Jasper, H., & Lehmann, R. (2011). Peroxiredoxin stabilization of DE-cadherin promotes primordial germ cell adhesion. *Developmental Cell, 20*, 233–243. https://doi.org/10.1016/j.devcel.2010.12.007.

Deshpande, G., Ng, C., Jourjine, N., Chiew, J. W., Dasilva, J., & Schedl, P. (2023). Hedgehog signaling guides migration of primordial germ cells to the Drosophila somatic gonad. *Genetics, 225*, iyad165. https://doi.org/10.1093/genetics/iyad165.

Devreotes, P., & Horwitz, A. R. (2015). Signaling networks that regulate cell migration. *Cold Spring Harbor Perspectives in Biology, 7*, a005959. https://doi.org/10.1101/cshperspect.a005959.

Doitsidou, M., Reichman-Fried, M., Stebler, J., Köprunner, M., Dörries, J., Meyer, D., ... Raz, E. (2002). Guidance of primordial germ cell migration by the chemokine SDF-1. *Cell, 111*, 647–659. https://doi.org/10.1016/S0092-8674(02)01135-2.

Elosegui-Artola, A., Trepat, X., & Roca-Cusachs, P. (2018). Control of mechanotransduction by molecular clutch dynamics. *Trends in Cell Biology, 28*, 356–367. https://doi.org/10.1016/j.tcb.2018.01.008.

Extavour, C. G., & Akam, M. (2003). Mechanisms of germ cell specification across the metazoans: Epigenesis and preformation. *Development (Cambridge, England), 130*, 5869–5884. https://doi.org/10.1242/dev.00804.

Friedl, P., Wolf, K., & Lammerding, J. (2011). Nuclear mechanics during cell migration. *Current Opinion in Cell Biology, Cell Structure and Dynamics, 23*, 55–64. https://doi.org/10.1016/j.ceb.2010.10.015.

Garcia-Arcos, J. M., Chabrier, R., Deygas, M., Nader, G., Barbier, L., Sáez, P. J., ... Piel, M. (2019). Reconstitution of cell migration at a glance. *Journal of Cell Science, 132*, jcs225565. https://doi.org/10.1242/jcs.225565.

García-Arcos, J. M., Jha, A., Waterman, C. M., & Piel, M. (2024). Blebology: Principles of bleb-based migration. *Trends in Cell Biology, 34*, 838–853. https://doi.org/10.1016/j.tcb.2024.02.009.

Georgantzoglou, A., Poplimont, H., Walker, H. A., Lämmermann, T., & Sarris, M. (2022). A two-step search and run response to gradients shapes leukocyte navigation in vivo. *Journal of Cell Biology, 221*, e202103207. https://doi.org/10.1083/jcb.202103207.

Godin, I., Deed, R., Cooke, J., Zsebo, K., Dexter, M., & Wylie, C. C. (1991). Effects of the steel gene product on mouse primordial germ cells in culture. *Nature, 352*, 807–809. https://doi.org/10.1038/352807a0.

Goudarzi, M., Banisch, T. U., Mobin, M. B., Maghelli, N., Tarbashevich, K., Strate, I., ... Raz, E. (2012). Identification and regulation of a molecular module for bleb-based cell motility. *Developmental Cell, 23*, 210–218. https://doi.org/10.1016/j.devcel.2012.05.007.

Grimaldi, C., & Raz, E. (2020). Germ cell migration—Evolutionary issues and current understanding. *Seminars in Cell & Developmental Biology, Regeneration of Vertebrate Organs edited by Jia-Qiang He. Cell migration edited by Michelle Starz-Gaiano, 100*, 152–159. https://doi.org/10.1016/j.semcdb.2019.11.015.

Grimaldi, C., Schumacher, I., Boquet-Pujadas, A., Tarbashevich, K., Vos, B. E., Bandemer, J., ... Raz, E. (2020). E-cadherin focuses protrusion formation at the front of migrating cells by impeding actin flow. *Nature Communications, 11*, 5397. https://doi.org/10.1038/s41467-020-19114-z.

Gross-Thebing, S., Truszkowski, L., Tenbrinck, D., Sánchez-Iranzo, H., Camelo, C., Westerich, K. J., ... Raz, E. (2020). Using migrating cells as probes to illuminate features in live embryonic tissues. *Science Advances, 6*, eabc5546. https://doi.org/10.1126/sciadv.abc5546.

Gross-Thebing, T., Yigit, S., Pfeiffer, J., Reichman-Fried, M., Bandemer, J., Ruckert, C., ... Raz, E. (2017). The vertebrate protein dead end maintains primordial germ cell fate by inhibiting somatic differentiation. *Developmental Cell, 43*, 704–715.e5. https://doi.org/10.1016/j.devcel.2017.11.019.

Gu, Y., Runyan, C., Shoemaker, A., Surani, A., & Wylie, C. (2009). Steel factor controls primordial germ cell survival and motility from the time of their specification in the allantois, and provides a continuous niche throughout their migration. *Development (Cambridge, England), 136*, 1295–1303. https://doi.org/10.1242/dev.030619.

Gu, Y., Runyan, C., Shoemaker, A., Surani, M. A., & Wylie, C. (2011). Membrane-bound steel factor maintains a high local concentration for mouse primordial germ cell motility, and defines the region of their migration. *PLoS One, 6*, e25984. https://doi.org/10.1371/journal.pone.0025984.

Hanyu-Nakamura, K., Kobayashi, S., & Nakamura, A. (2004). Germ cell-autonomous Wunen2 is required for germline development in Drosophila embryos. *Development (Cambridge, England), 131*, 4545–4553. https://doi.org/10.1242/dev.01321.

Hartwig, J., Tarbashevich, K., Seggewiß, J., Stehling, M., Bandemer, J., Grimaldi, C., ... Raz, E. (2014). Temporal control over the initiation of cell motility by a regulator of G-protein signaling. *Proceedings of the National Academy of Sciences, 111*, 11389–11394. https://doi.org/10.1073/pnas.1400043111.

Hawkins, R. J., Poincloux, R., Bénichou, O., Piel, M., Chavrier, P., & Voituriez, R. (2011). Spontaneous contractility-mediated cortical flow generates cell migration in three-dimensional environments. *Biophysical Journal, 101*, 1041–1045. https://doi.org/10.1016/j.bpj.2011.07.038.

Henson, J. H., Svitkina, T. M., Burns, A. R., Hughes, H. E., MacPartland, K. J., Nazarian, R., & Borisy, G. G. (1999). Two components of actin-based retrograde flow in sea urchin coelomocytes. *MBoC, 10*, 4075–4090. https://doi.org/10.1091/mbc.10.12.4075.

Hinck, L. (2004). The versatile roles of "axon guidance" cues in tissue morphogenesis. *Developmental Cell, 7*, 783–793. https://doi.org/10.1016/j.devcel.2004.11.002.

Hind, L. E., Vincent, W. J. B., & Huttenlocher, A. (2016). Leading from the back: The role of the uropod in neutrophil polarization and migration. *Developmental Cell, 38*, 161–169. https://doi.org/10.1016/j.devcel.2016.06.031.

Hughes, C. E., & Nibbs, R. J. B. (2018). A guide to chemokines and their receptors. *The FEBS Journal, 285*, 2944–2971. https://doi.org/10.1111/febs.14466.

Jaglarz, M. K., & Howard, K. R. (1995). The active migration of Drosophila primordial germ cells. *Development (Cambridge, England), 121*, 3495–3503. https://doi.org/10.1242/dev.121.11.3495.

Jarysta, A., Riou, L., Firlej, V., Lapoujade, C., Kortulewski, T., Barroca, V., ... Fouchet, P. (2022). Abnormal migration behavior linked to Rac1 signaling contributes to primordial germ cell exhaustion in Fanconi anemia pathway-deficient Fancg−/− embryos. *Human Molecular Genetics, 31*, 97–110. https://doi.org/10.1093/hmg/ddab222.

Kardash, E., Reichman-Fried, M., Maître, J.-L., Boldajipour, B., Papusheva, E., Messerschmidt, E.-M., ... Raz, E. (2010). A role for Rho GTPases and cell–cell adhesion in single-cell motility in vivo. *Nature Cell Biology, 12*, 47–53. https://doi.org/10.1038/ncb2003.

Kim, J. H., Hanlon, C. D., Vohra, S., Devreotes, P. N., & Andrew, D. J. (2021). Hedgehog signaling and Tre1 regulate actin dynamics through PI(4,5)P2 to direct migration of Drosophila embryonic germ cells. *Cell Reports, 34*, 108799. https://doi.org/10.1016/j.celrep.2021.108799.

Kirschner, M., & Gerhart, J. (1998). Evolvability. *Proceedings of the National Academy of Sciences, 95*, 8420–8427. https://doi.org/10.1073/pnas.95.15.8420.

Knaut, H., Werz, C., Geisler, R., & Nüsslein-Volhard, C. (2003). A zebrafish homologue of the chemokine receptor Cxcr4 is a germ-cell guidance receptor. *Nature, 421*, 279–282. https://doi.org/10.1038/nature01338.

Krieg, M., Arboleda-Estudillo, Y., Puech, P.-H., Käfer, J., Graner, F., Müller, D. J., & Heisenberg, C.-P. (2008). Tensile forces govern germ-layer organization in zebrafish. *Nature Cell Biology, 10*, 429–436. https://doi.org/10.1038/ncb1705.

Kunwar, P. S., Sano, H., Renault, A. D., Barbosa, V., Fuse, N., & Lehmann, R. (2008). Tre1 GPCR initiates germ cell transepithelial migration by regulating Drosophila melanogaster E-cadherin. *Journal of Cell Biology, 183*, 157–168. https://doi.org/10.1083/jcb.200807049.

Kunwar, P. S., Starz-Gaiano, M., Bainton, R. J., Heberlein, U., & Lehmann, R. (2003). Tre1, a G protein-coupled receptor, directs transepithelial migration of drosophila germ cells. *PLoS Biology, 1*, e80. https://doi.org/10.1371/journal.pbio.0000080.

Lämmermann, T., & Sixt, M. (2009). Mechanical modes of 'amoeboid' cell migration. *Current Opinion in Cell Biology, 21*, 636–644. https://doi.org/10.1016/j.ceb.2009.05.003.

LeBlanc, M. G., & Lehmann, R. (2017). Domain-specific control of germ cell polarity and migration by multifunction Tre1 GPCR. *Journal of Cell Biology, 216*, 2945–2958. https://doi.org/10.1083/jcb.201612053.

Lin, B., Luo, J., & Lehmann, R. (2022). An AMPK phosphoregulated RhoGEF feedback loop tunes cortical flow–driven amoeboid migration in vivo. *Science Advances, 8*, eabo0323. https://doi.org/10.1126/sciadv.abo0323.

Malhotra, D., Shin, J., Solnica-Krezel, L., & Raz, E. (2018). Spatio-temporal regulation of concurrent developmental processes by generic signaling downstream of chemokine receptors. *eLife, 7*, e33574. https://doi.org/10.7554/eLife.33574.

McGregor, A. L., Hsia, C.-R., & Lammerding, J. (2016). Squish and squeeze—The nucleus as a physical barrier during migration in confined environments. *Current Opinion in Cell Biology, Cell nucleus, 40*, 32–40. https://doi.org/10.1016/j.ceb.2016.01.011.

Meyen, D., Tarbashevich, K., Banisch, T. U., Wittwer, C., Reichman-Fried, M., Maugis, B., ... Raz, E. (2015). Dynamic filopodia are required for chemokine-dependent intracellular polarization during guided cell migration in vivo. *eLife, 4*, e05279. https://doi.org/10.7554/eLife.05279.

Minina, S., Reichman-Fried, M., & Raz, E. (2007). Control of receptor internalization, signaling level, and precise arrival at the target in guided cell migration. *Current Biology, 17*, 1164–1172. https://doi.org/10.1016/j.cub.2007.05.073.

Mitchison, T., & Kirschner, M. (1988). Cytoskeletal dynamics and nerve growth. *Neuron, 1*, 761–772. https://doi.org/10.1016/0896-6273(88)90124-9.

Molyneaux, K. A., Stallock, J., Schaible, K., & Wylie, C. (2001). Time-lapse analysis of living mouse germ cell migration. *Developmental Biology, 240*, 488–498. https://doi.org/10.1006/dbio.2001.0436.

Molyneaux, K. A., Zinszner, H., Kunwar, P. S., Schaible, K., Stebler, J., Sunshine, M. J., ... Lehmann, R. (2003). The chemokine SDF1/CXCL12 and its receptor CXCR4 regulate mouse germ cell migration and survival. *Development (Cambridge, England), 130*, 4279–4286. https://doi.org/10.1242/dev.00640.

Moore, L. A., Broihier, H. T., Doren, M. V., Lunsford, L. B., & Lehmann, R. (1998). Identification of genes controlling germ cell migration and embryonic gonad formation in Drosophila. *Development (Cambridge, England), 125*, 667–678. https://doi.org/10.1242/dev.125.4.667.

Morita, M., Morimoto, M., Teramoto, T., Ikenouchi, J., Atsuta, Y., Saito, D. (2023). The SOCE system is critical for membrane bleb formation to drive avian primordial germ cell migration. https://doi.org/10.1101/2023.06.12.544577.

Murphy, J. W., Cho, Y., Sachpatzidis, A., Fan, C., Hodsdon, M. E., & Lolis, E. (2007). Structural and functional basis of CXCL12 (stromal cell-derived factor-1α) binding to heparin. *Journal of Biological Chemistry, 282*, 10018–10027. https://doi.org/10.1074/jbc.M608796200.

Nourshargh, S., & Alon, R. (2014). Leukocyte migration into inflamed tissues. *Immunity, 41*, 694–707. https://doi.org/10.1016/j.immuni.2014.10.008.

Olguin-Olguin, A., Aalto, A., Maugis, B., Boquet-Pujadas, A., Hoffmann, D., Ermlich, L., ... Raz, E. (2021). Chemokine-biased robust self-organizing polarization of migrating cells in vivo. *Proceedings of the National Academy of Sciences, 118*, e2018480118. https://doi.org/10.1073/pnas.2018480118.

Oosterhuis, J. W., & Looijenga, L. H. J. (2019). Human germ cell tumours from a developmental perspective. *Nature Reviews. Cancer, 19*, 522–537. https://doi.org/10.1038/s41568-019-0178-9.

Paksa, A., Bandemer, J., Hoeckendorf, B., Razin, N., Tarbashevich, K., Minina, S., ... Raz, E. (2016). Repulsive cues combined with physical barriers and cell–cell adhesion determine progenitor cell positioning during organogenesis. *Nature Communications, 7*, 11288. https://doi.org/10.1038/ncomms11288.

Paluch, E., Piel, M., Prost, J., Bornens, M., & Sykes, C. (2005). Cortical actomyosin breakage triggers shape oscillations in cells and cell fragments. *Biophysical Journal, 89*, 724–733. https://doi.org/10.1529/biophysj.105.060590.

Paluch, E. K., Aspalter, I. M., & Sixt, M. (2016). Focal adhesion–independent cell migration. *Annual Review of Cell and Developmental Biology, 32*, 469–490. https://doi.org/10.1146/annurev-cellbio-111315-125341.

Petrie, R. J., Doyle, A. D., & Yamada, K. M. (2009). Random versus directionally persistent cell migration. *Nature Reviews. Molecular Cell Biology, 10*, 538–549. https://doi.org/10.1038/nrm2729.

Pollard, T. D., Blanchoin, L., & Mullins, R. D. (2000). Molecular mechanisms controlling actin filament dynamics in nonmuscle cells. *Annual Review of Biophysics, 29*, 545–576. https://doi.org/10.1146/annurev.biophys.29.1.545.

Pyne, S., Kong, K.-C., & Darroch, P. I. (2004). Lysophosphatidic acid and sphingosine 1-phosphate biology: The role of lipid phosphate phosphatases. *Seminars in Cell & Developmental Biology, Biology of Lysophosphatidic Acid and Sphingosine 1-Phosphate and Polarity in Mammalian Development, 15*, 491–501. https://doi.org/10.1016/j.semcdb.2004.05.007.

Reichman-Fried, M., Minina, S., & Raz, E. (2004). Autonomous modes of behavior in primordial germ cell migration. *Developmental Cell, 6*, 589–596. https://doi.org/10.1016/S1534-5807(04)00074-7.

Renault, A. D., Kunwar, P. S., & Lehmann, R. (2010). Lipid phosphate phosphatase activity regulates dispersal and bilateral sorting of embryonic germ cells in Drosophila. *Development (Cambridge, England), 137*, 1815–1823. https://doi.org/10.1242/dev.046110.

Renault, A. D., & Lehmann, R. (2006). Follow the fatty brick road: Lipid signaling in cell migration. *Current Opinion in Genetics & Development, Pattern Formation and Developmental Mechanisms, 16*, 348–354. https://doi.org/10.1016/j.gde.2006.06.015.

Renault, A. D., Sigal, Y. J., Morris, A. J., & Lehmann, R. (2004). Soma-germ line competition for lipid phosphate uptake regulates germ cell migration and survival. *Science (New York, N. Y.), 305*, 1963–1966. https://doi.org/10.1126/science.1102421.

Richardson, B. E., & Lehmann, R. (2010). Mechanisms guiding primordial germ cell migration: Strategies from different organisms. *Nature Reviews. Molecular Cell Biology, 11*, 37–49. https://doi.org/10.1038/nrm2815.

Ridley, A. J. (2001). Rho GTPases and cell migration. *Journal of Cell Science, 114*, 2713–2722. https://doi.org/10.1242/jcs.114.15.2713.

Ruprecht, V., Wieser, S., Callan-Jones, A., Smutny, M., Morita, H., Sako, K., ... Heisenberg, C.-P. (2015). Cortical contractility triggers a stochastic switch to fast amoeboid cell motility. *Cell, 160*, 673–685. https://doi.org/10.1016/j.cell.2015.01.008.

Schick, J., & Raz, E. (2022). Blebs—Formation, regulation, positioning, and role in amoeboid cell migration. *Frontiers in Cell and Developmental Biology, 10*.

Sedzinski, J., Biro, M., Oswald, A., Tinevez, J.-Y., Salbreux, G., & Paluch, E. (2011). Polar actomyosin contractility destabilizes the position of the cytokinetic furrow. *Nature, 476*, 462–466. https://doi.org/10.1038/nature10286.

Seifert, J. R. K., & Lehmann, R. (2012). Drosophila primordial germ cell migration requires epithelial remodeling of the endoderm. *Development (Cambridge, England), 139*, 2101–2106. https://doi.org/10.1242/dev.078949.

Starz-Gaiano, M., Cho, N. K., Forbes, A., & Lehmann, R. (2001). Spatially restricted activity of a Drosophila lipid phosphatase guides migrating germ cells. *Development (Cambridge, England), 128*, 983–991. https://doi.org/10.1242/dev.128.6.983.

Stebler, J., Spieler, D., Slanchev, K., Molyneaux, K. A., Richter, U., Cojocaru, V., ... Raz, E. (2004). Primordial germ cell migration in the chick and mouse embryo: The role of the chemokine SDF-1/CXCL12. *Developmental Biology, 272*, 351–361. https://doi.org/10.1016/j.ydbio.2004.05.009.

Szabó, A., & Mayor, R. (2018). Mechanisms of neural crest migration. *Annual Review of Genetics, 52*, 43–63. https://doi.org/10.1146/annurev-genet-120417-031559.

Takeuchi, T., Tanigawa, Y., Minamide, R., Ikenishi, K., & Komiya, T. (2010). Analysis of SDF-1/CXCR4 signaling in primordial germ cell migration and survival or differentiation in *Xenopus laevis*. *Mechanisms of Development, 127*, 146–158. https://doi.org/10.1016/j.mod.2009.09.005.

Tarbashevich, K., Dzementsei, A., & Pieler, T. (2011). A novel function for KIF13B in germ cell migration. *Developmental Biology, 349*, 169–178. https://doi.org/10.1016/j.ydbio.2010.10.016.

Terayama, K., Kataoka, K., Morichika, K., Orii, H., Watanabe, K., & Mochii, M. (2013). Developmental regulation of locomotive activity in enopus primordial germ cells. *Development, Growth & Differentiation, 55*, 217–228. https://doi.org/10.1111/dgd.12018.

Thorpe, J. L., Doitsidou, M., Ho, S.-Y., Raz, E., & Farber, S. A. (2004). Germ cell migration in zebrafish is dependent on HMGCoA reductase activity and prenylation. *Developmental Cell, 6*, 295–302. https://doi.org/10.1016/S1534-5807(04)00032-2.

Trappmann, B., & Chen, C. S. (2013). How cells sense extracellular matrix stiffness: A material's perspective. *Current Opinion in Biotechnology, Tissue, Cell and Engineering, 24*, 948–953. https://doi.org/10.1016/j.copbio.2013.03.020.

Truszkowski, L., Batur, D., Long, H., Tarbashevich, K., Vos, B. E., Trappmann, B., & Raz, E. (2023). Primordial germ cells adjust their protrusion type while migrating in different tissue contexts in vivo. *Development (Cambridge, England), 150*, dev200603. https://doi.org/10.1242/dev.200603.

Van Doren, M., Broihier, H. T., Moore, L. A., & Lehmann, R. (1998). HMG-CoA reductase guides migrating primordial germ cells. *Nature, 396*, 466–469. https://doi.org/10.1038/24871.

Wadhwa, N., & Berg, H. C. (2022). Bacterial motility: Machinery and mechanisms. *Nature Reviews. Microbiology, 20*, 161–173. https://doi.org/10.1038/s41579-021-00626-4.

Warrior, R. (1994). Primordial germ cell migration and the assembly of the drosophila embryonic gonad. *Developmental Biology, 166*, 180–194. https://doi.org/10.1006/dbio.1994.1306.

Weidinger, G., Wolke, U., Köprunner, M., Klinger, M., & Raz, E. (1999). Identification of tissues and patterning events required for distinct steps in early migration of zebrafish primordial germ cells. *Development (Cambridge, England), 126*, 5295–5307. https://doi.org/10.1242/dev.126.23.5295.

Weidinger, G., Wolke, U., Köprunner, M., Thisse, C., Thisse, B., & Raz, E. (2002). Regulation of zebrafish primordial germ cell migration by attraction towards an intermediate target. *Development (Cambridge, England), 129*, 25–36. https://doi.org/10.1242/dev.129.1.25.

Westerich, K. J., Tarbashevich, K., Schick, J., Gupta, A., Zhu, M., Hull, K., ... Raz, E. (2023). Spatial organization and function of RNA molecules within phase-separated condensates in zebrafish are controlled by Dnd1. *Developmental Cell, 58*, 1578–1592.e5. https://doi.org/10.1016/j.devcel.2023.06.009.

Yoon, C., Kawakami, K., & Hopkins, N. (1997). Zebrafish vasa homologue RNA is localized to the cleavage planes of 2- and 4-cell-stage embryos and is expressed in the primordial germ cells. *Development (Cambridge, England), 124*, 3157–3165. https://doi.org/10.1242/dev.124.16.3157.

Zhang, N., Zhang, J., Purcell, K. J., Cheng, Y., & Howard, K. (1997). The Drosophila protein Wunen repels migrating germ cells. *Nature, 385*, 64–67. https://doi.org/10.1038/385064a0.

Zhang, Y., Wei, D., Wang, X., Wang, B., Li, M., Fang, H., ... Ye, F. (2023). Run-and-tumble dynamics and mechanotaxis discovered in microglial migration. *Research; A Journal of Science and Its Applications, 6*, 0063. https://doi.org/10.34133/research.0063.

Zlotnik, A., & Yoshie, O. (2012). The chemokine superfamily revisited. *Immunity, 36*, 705–716. https://doi.org/10.1016/j.immuni.2012.05.008.

CHAPTER TWO

Outside influences: The impact of extracellular matrix mechanics on cell migration

Ronen Zaidel-Bar[a,*] **and Priti Agarwal**[a,b,*]
[a]Department of Cell and Developmental Biology, School of Medicine, Faculty of Medical and Health Sciences, Tel Aviv University, Tel Aviv, Israel
[b]National Center for Biological Sciences, Tata Institute for Fundamental Research, Bangalore, Karnataka, India
*Corresponding authors. e-mail address: Zaidelbar@tauex.tau.ac.il; pritiag007@gmail.com

Contents

1. Introduction 30
2. ECM composition and structure 30
   2.1 Interstitial matrix 31
   2.2 Basement membrane 33
3. Modulators of ECM mechanics 34
   3.1 Cross-linking enzymes 35
   3.2 Matrix proteases 35
4. Mechanical role of ECM in cell migration 36
   4.1 ECM fibrillar architecture 36
   4.2 ECM porosity 42
   4.3 ECM dynamics 45
   4.4 Cell-matrix anchorage 47
   4.5 ECM Stiffness 49
5. Outlook 52
References 55

Abstract

"No cell is an island" – highlights the interconnectedness of cellular behavior and the extracellular matrix (ECM). Cell migration is inherently contextual, as cells navigate and adapt to their environments, reshaping the ECM while being influenced by its properties. This review focuses on the mechanical characteristics of the ECM—specifically its architecture, porosity, dynamics, and stiffness—and how these attributes affect cell behavior and migration strategies. We discuss how the mechanical properties are modulated by the composition and arrangement of ECM components and the role of enzymatic activities, including crosslinking and matrix metalloproteinases. By presenting examples from vertebrate and invertebrate developmental models, we demonstrate how ECM mechanics dictate cell migration at various biological scales. Additionally, we examine the importance of cell-matrix adhesions in regulating migration speed and

Current Topics in Developmental Biology, Volume 164
ISSN 0070-2153, https://doi.org/10.1016/bs.ctdb.2025.01.003

direction. While in vitro studies have advanced our understanding of the molecular mechanisms at play, significant questions persist regarding the regulation of cell migration by ECM mechanics in vivo. Ultimately, this synthesis aims to illuminate the complexities of cell-ECM mechanical interactions, pointing the way for future research that may unveil novel insights into how ECM mechanics influences cell migration during development and disease.

1. Introduction

"Traveler, there is no path. The path is made by walking." — Antonio Machado
"The journey itself is what transforms you." — Rumi

Cell migration within the animal body exemplifies a dynamic relationship between cells and their environment, where each shapes the other. As cells traverse through different regions, they modify the extracellular matrix (ECM) to forge new paths, while the ECM's biochemical and mechanical properties, in turn, influence cell behavior and migration strategies. The biochemical aspects of the extracellular matrix affecting cell migration have been covered elsewhere (Pally & Naba, 2024; Scarpa & Mayor, 2016; Sherwood, 2021). Here, we focus on the mechanical properties of the matrix, including architecture, porosity, dynamics, and stiffness, and explain how they are modulated by the composition and organization of components, as well as enzymatic modifications such as crosslinking and matrix metalloprotease activity (Naba, 2024). We highlight examples from vertebrate and invertebrate developmental models that demonstrate how ECM mechanics govern cell migration at both cell and tissue scales (Pally & Naba, 2024; Ventura & Sedzinski, 2022). Additionally, we discuss various roles of cell-matrix adhesions in regulating cell migration. In vitro studies provide insight into the molecular mechanisms underlying some of these processes (Agarwal & Zaidel-Bar, 2019; Le & Mayor, 2023; Yamada & Sixt, 2019), but many questions remain open with regard to the regulation of cell migration by ECM mechanics in vivo. At the end of the chapter we point out the challenges and opportunities in capturing cell-ECM interactions in vivo.

2. ECM composition and structure

The ECM is a complex meshwork organized in two main forms: interstitial matrix and basement membrane (BM)(Walma & Yamada, 2020; Frantz et al., 2010). The interstitial matrix is a three-dimensional highly-hydrated porous network surrounding mesenchymal cells and contributing to the volume of tissues, while BM is a tightly-woven thin sheet-like matrix that underlies epithelial cells and surrounds tissues (Fig. 1).

2.1 Interstitial matrix

The composition of ECM is amazingly diverse with many interwoven components, including collagens, fibronectins, lamins, proteoglycans, and hyaluronic acid, to name but a few. They contribute to the various mechanical properties of the extracellular environment such as stiffness, elasticity, viscoelasticity, porosity etc. (Saraswathibhatla, Indana, & Chaudhuri, 2023; Yamada & Sixt, 2019; Yamada et al., 2022) (Fig. 2A–F). Below we discuss a few of the major ECM components and how they contribute to ECM mechanics.

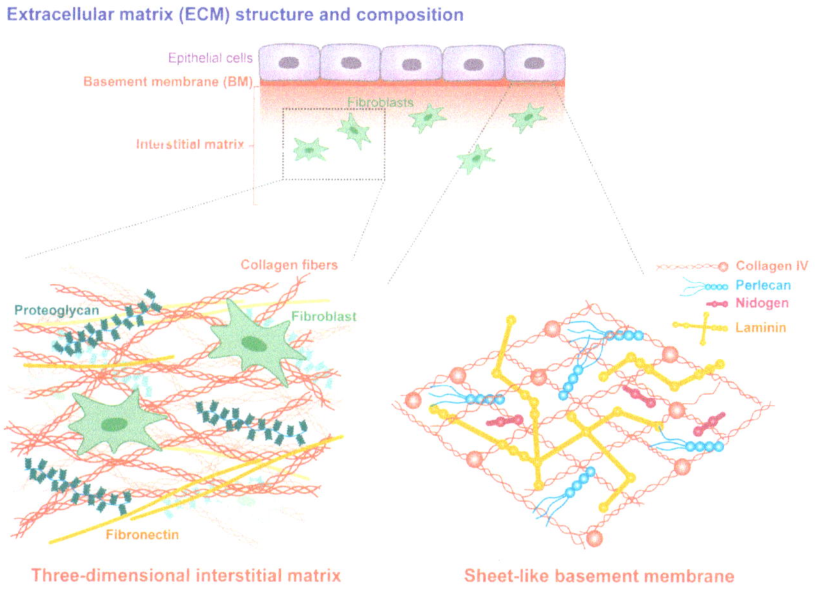

Fig. 1 Structure and composition of the extracellular matrix. The extracellular matrix (ECM) exists in two forms: the interstitial matrix and the basement membrane (BM). These matrices are structurally distinct, each composed of specific ECM proteins. The interstitial matrix is a three-dimensional fibrillar network that fills the spaces between cells within organs and is primarily enriched in collagen fibers, fibronectin, and proteoglycans. In contrast, the BM forms a thin, two-dimensional sheet that underlies epithelial cells and surrounds most of our organs and is mainly composed of laminin, collagen IV, perlecan, and nidogen.

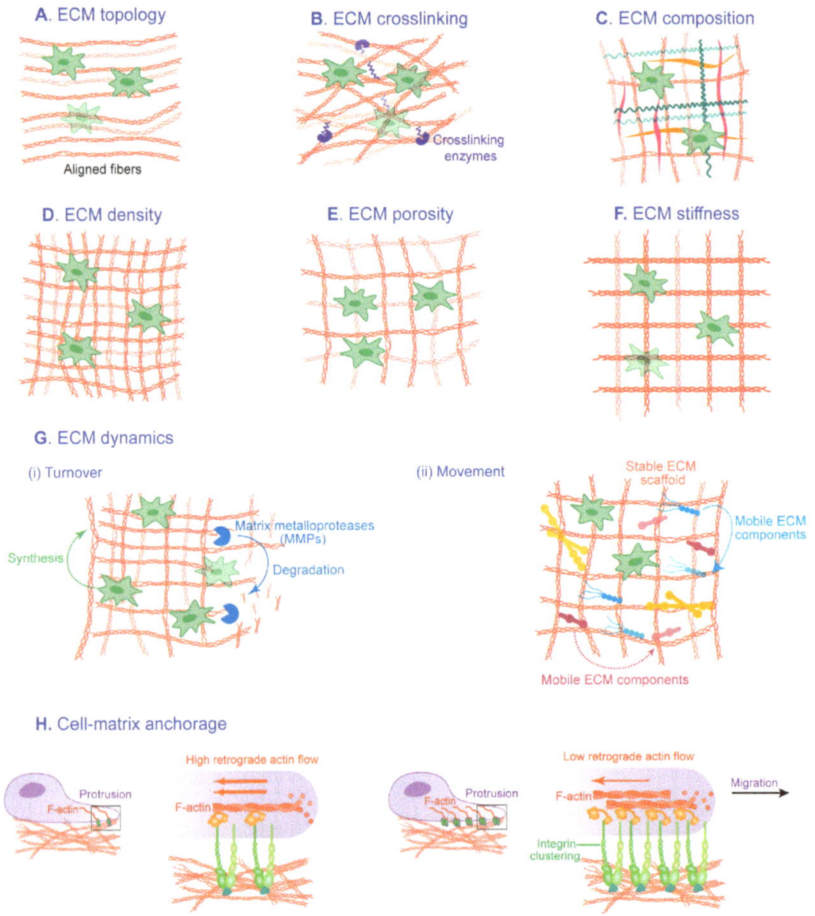

Fig. 2 Mechanical properties of the extracellular matrix influencing cell migration. Multiple aspects of ECM mechanics influence cell migration. The aspects discussed here include (A) ECM topology, (B) ECM crosslinking, (C) ECM composition, (D) ECM density, (E) ECM porosity, (F) ECM stiffness, (G) ECM dynamics, and (H) Cell-matrix anchorage.

Collagen-I: It is one of the major ECM components comprising nearly 90% of the human ECM. Secreted by fibroblasts, osteoblasts, and odontoblasts, collagen-I has a triple helical structure with two α1 and one α2 chain. Assembling into higher order structures, collagen fibrils form fibers with a thickness ranging from 50 nm to 100 nm, arranged in networks (Ricard-Blum, 2011). The organization of collagen networks is highly tissue specific, and they contribute to several mechanical properties of the matrix including stiffness, nonlinear elasticity, plasticity, and viscoelasticity

(Ban et al., 2018; Münster et al., 2013; Nam, Hu, Butte, & Chaudhuri, 2016). Collagen fibers can be covalently crosslinked by enzymes from the lysl oxidase family, which further increases collagen network density and stiffness (Barker, Cox, & Erler, 2012).

Fibronectin: Fibronectin (FN) is a complex glycoprotein that interacts with various ECM components, such as heparan sulfate, collagen, fibrin and other fibronectin molecules, and growth factors (Dalton & Lemmon, 2021; Geiger & Yamada, 2011; Pankov & Yamada, 2002). After initial clustering that is oxidation-dependent (Melamed et al., 2023), FN fibrillogenesis is a force-dependent process that depends on its interaction with cell-surface receptors. Soluble FN fibers assemble into an insoluble fibrillar matrix through the interaction of its two arginine-glycine-aspartate (RGD) domains with the β1- and β3-containing cell-surface receptor integrins (Humphries et al., 2006). Pulling forces applied to FN open cryptic domains that subsequently allow its crosslinking with integrins and other ECM molecules to form a fibrillar matrix (Baneyx, Baugh, & Vogel, 2002; Huet-Calderwood et al., 2023).

Although the mechanical properties of interconnected FN fibers are not yet fully understood, computational models and in vitro studies have shown that FN fibers are exceptionally extensible and can return to their original form once mechanical force is removed (Klotzsch et al., 2009; Little et al., 2008; Ohashi, Kiehart, & Erickson, 1999). Since FN fibrils bind to various ECM molecules, the extent to which these fibrils are stretched can alter their binding capacity and the number of available binding sites for other ECM proteins. This dynamic interaction, in turn, also affects the mechanical properties of the FN fibrils (Bradshaw, Hoffmann, Wong, & Smith, 2019).

Glycosaminoglycans (GAGs): These are unbranched polysaccharides made up of repeating units of specific disaccharides. In vertebrates, the main GAGs are: heparan sulfate, chondroitin sulfate, keratan sulfate and hyaluronan. GAGs are highly negatively charged and that attracts cations and causes an influx of water into the matrix, thus forming hydrogels. The increase in osmotic pressure enables the matrix to endure compressive forces (Scott, 2003). Thus, GAGs impart the matrix with compressive and tensile strength. Also, in some cases they modify the porosity of the ECM (Chaudhari et al., 2020).

2.2 Basement membrane

The BM is a thin laminar matrix enwrapping most of the organs in a multicellular organism. During development, the BM performs a plethora of functions: it is a storehouse of many signaling molecules, regulating polarity, cell fate specification, and cell migration (Pozzi, Yurchenco, & Iozzo, 2017; Sekiguchi & Yamada, 2018; Sherwood, 2021). BMs also have several structural roles, including barrier function, compartmentalization and connecting tissues (Brown, 2011; Jayadev & Sherwood, 2017; Morrissey et al., 2014; Yurchenco, 2011). Recently, several studies have revealed the dynamic and mechanical role of BMs in shaping several tissues and organs. These findings have been comprehensively discussed in several excellent reviews (Díaz-de-la-Loza & Stramer, 2024; Sherwood, 2021; Töpfer, 2023; Wu et al., 2023).

The BM is mainly composed of four conserved proteins: laminin, collagen IV, nidogen and perlecan. The glycoprotein laminin is a cross-shaped heterotrimer composed of α, β, and γ chains (Yurchenco, 2011). Laminin initiates the assembly of BM. The N-terminal domain of laminin forms a network through interactions with α, β, or γ subunits of adjacent laminin heterotrimers while the laminin globular domain binds to cell surface receptors like integrins and dystroglycans. Following laminin polymerization and assembly, there is a hierarchical recruitment of other BM components: collagen IV, nidogen, and perlecan (Keeley et al., 2020; Matsubayashi et al., 2017). Collagen provides stiffness and tensile strength to the BM (Fidler et al., 2018). Laminin and collagen networks are further crosslinked by the glycoprotein nidogen and proteoglycan perlecan to form a polymeric network. BM also includes many other components such as agrin, fibulin-1, the glycoprotein SPARC and others (Aumailley, Wiedemann, Mann, & Timpl, 1989; Fox et al., 1991; Hopf et al., 2001). BM composition and organization dictate several mechanical properties of the BM such as stiffness, tensile strength, porosity, elasticity, and its viscoelastic nature (Elosegui-Artola, 2021; Khalilgharibi & Mao, 2021). It is also important to remember that an alteration in BM composition or organization can affect multiple mechanical properties simultaneously. For example, an increase in collagen level can affect the matrix porosity as well as its stiffness.

3. Modulators of ECM mechanics

The mechanical nature of the ECM is determined by the composition, concentration, and connectivity of various ECM components (some of which were discussed in the previous section). Additionally, ECM mechanics is modulated by cell-secreted enzymes, such as crosslinkers and matrix metalloproteases.

3.1 Cross-linking enzymes

Matrix crosslinking, mediated by enzymes such as lysyl oxidases (LOX) and transglutaminases, plays a critical role in modulating the mechanical properties of the ECM. Typically, enhanced cross-linking leads to increased accumulation of ECM and stiffening of the matrix (Piersma, Hayward, & Weaver, 2020). Matrix cross-linking also reduces its turnover (Filipe et al., 2018), and the coexistence of weak and strong crosslinks within the matrix render it viscoelastic in nature (Nam et al., 2016). LOX family enzymes are required for the crosslinking of collagen and elastin. They cause oxidative deamination of lysyl and hydroxyl residues, forming reactive aldehydes. These aldehydes further react with nearby lysyl and hydroxyl residues and thus, covalently crosslink the matrix. Abberant LOX-mediated ECM remodeling is associated with fibrosis and facillitates tumor invasion and metastasis (Barker et al., 2012). Transglutaminases are involved in transamidation and deamidation of glutamine and asparagine residues of various ECM proteins, such as collagen IV (Cardoso et al., 2015), fibrinogen (Barsigian, Fellin, Jain, & Martinez, 1988), and proteoglycans (Scarpellini et al., 2009). The action of transglutaminase leads to alligmenment of matrix fibers during ECM synthesis (Espitia Pinzón et al., 2017), ECM stiffening (Spurlin et al., 2009) and reduced proteolysis (Larreta-Garde & Berry, 2002). Similar to LOX enzymes, transglutaminases promote migration and cancer cell invasion (Lee et al., 2018).

3.2 Matrix proteases

Matrix degradation and turnover is critical for tissue homeostasis and is mediated by zinc-containing endopeptidases including Matrix MetalloProteases (MMPs) and A Disintegrin And Metalloproteinase proteins with ThromboSpondin motifs (ADAMTSs). Humans have a total of 23 MMPs and they all share a hemopexin domain, a Zn^{2+}-binding catalytic domain, and a self-inhibitory prodomain at the N-terminus. MMPs impact matrix integrity and several cellular processes by degrading matrix proteins and via the release of cell-surface bound receptors, cytokines and growth factors (Dzobo & Dandara, 2023; Shimoda et al., 2021). MMPs play a crucial role in modulating various physical features of the ECM, including its overall architecture, porosity, and dynamic remodeling (see Sections 4.1, 4.2 and 4.3). The ADAMTS proteinase family includes 19 members in human and the common structural domains includes a prodomain, the metalloproteinase, and disintegrin domain. They are secreted enzymes and most of them are involved in degrading proteoglycans like aggrecan, versican, and brevican (Apte, 2009). ADAMTS-2 processes procollagen and hence is also responsible for depositing collagen fibrils (Bonnans, Chou, & Werb, 2014).

4. Mechanical role of ECM in cell migration

The ECM is actively and dynamically formed and remodeled, imparting each tissue with a unique matrix composition, organization and topology. As migrating cells navigate across and through native tissue environments, the ECM they encounter continuously changes, like a traveler encountering different terrains. How exactly the ultrastructure and layout of the ECM guide these migrating cells in vivo is only just beginning to unravel (Figs. 3 and 4). In the following sections, we review recent findings that demonstrate how various mechanical aspects of the ECM affect cell migration. While an increasing amount of work is being performed in vivo, insight into the molecular mechanisms linking ECM mechanics to cellular behavior often stems from in vitro experiments.

4.1 ECM fibrillar architecture

Migrating cells actively secrete and reorganize the ECM to facilitate directional cell movement. A recent study in chick embryos by Martinson et al. (2023), shed light on how the deposition and remodeling of immature punctate FN by migrating cranial neural crest cells (NCCs) promotes their long-range collective cell migration (Martinson et al., 2023). "Leader" neural crest cells remodel an initially unorganized and punctate FN deposition into linear streamlined fibrils. Using both experiments and simulations, the authors showed that these fibrils act as guidance cues for the "follower" NCCs, steering them into tracks of coherent streams. In the absence of FN remodeling by the leader cells, migrating NCCs enter a "jammed" state, where the cells are densely packed (Sadati, Taheri Qazvini, Krishnan, Park, & Fredberg, 2013), and the migration efficiency is significantly decreased. It is fascinating to note that only leader cells play a crucial role in this ECM remodeling, while the follower cells do not. Even when the follower cells are unable to remodel FN, collective cell migration can continue, as long as the leaders are actively doing their part. Thus, the leader and follower NCCs have specialized roles in ECM remodeling, enabling robust long-distance collective cell migration by preventing cell jamming.

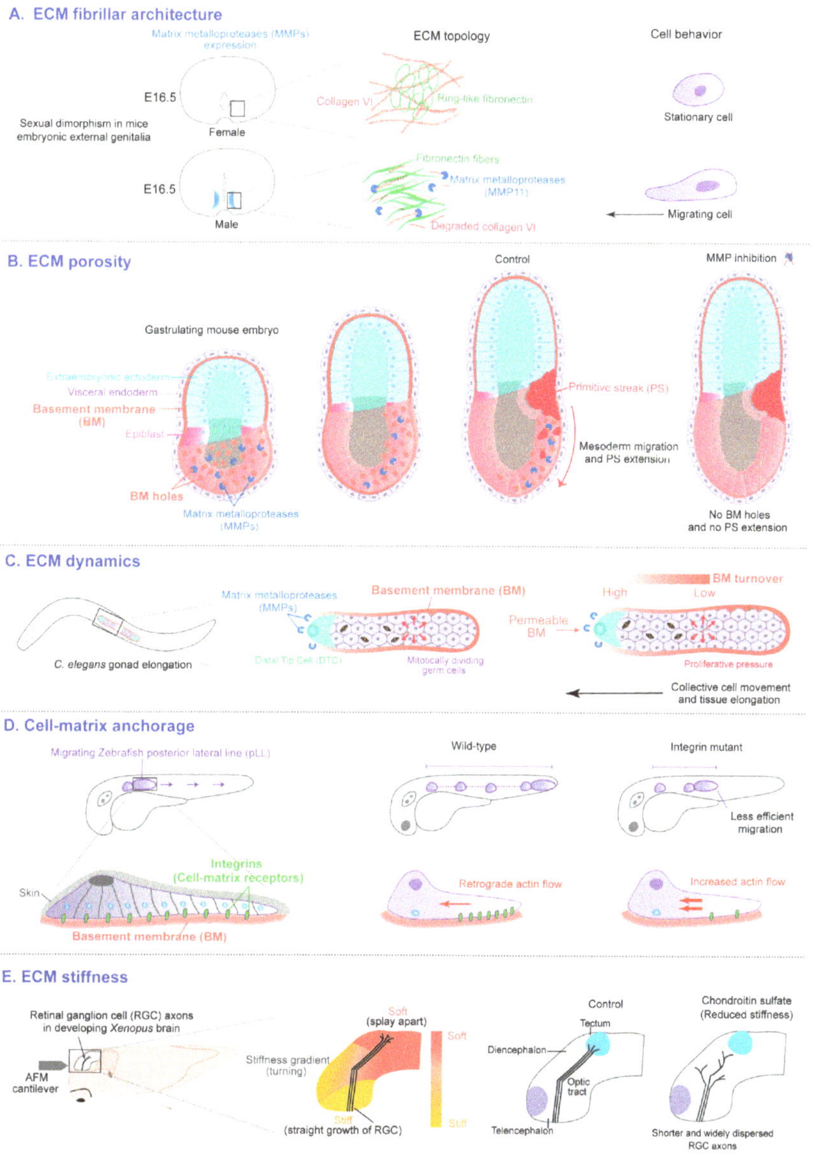

Fig. 3 A few in vivo examples depicting the role of ECM mechanics in cell migration. (A) **ECM architecture:** During sexual differentiation of embryonic mouse genitalia, remodeling of the ECM architecture, driven by the matrix metalloprotease MMP11 (blue), stimulates cell migration. At embryonic day 16.5 (E16.5), MMP11 expression is significantly higher in male than in female external genitalia. In female embryos, the ECM has collagen VI and a ring-like pattern of fibronectin (FN) (green). In males, however, MMP11 cleaves collagen VI, leading to the formation of fibrillar FN. This reorganization of the ECM induces cell migration, contributing to the

(Continued)

Similar to fibronectin fibers, cross-linked collagen fibers stimulate directional cell migration. For example, in *Zebrafish*, when part of the tail fin is amputated, thick bundles of collagen fibers form at the site of injury. These fibers serve as directional cues for the migration of endothelial cells, facilitating proper angiogenesis, as necessary for fin regeneration. Treatment with BAPN, an inhibitor of collagen cross-linking, disrupts endothelial cell migration, leading to impaired blood vessel formation and compromised tissue regeneration (Senk & Djonov, 2021).

Fibrillar ECM topology is modulated by the interaction of multiple ECM proteins, by cell-surface receptors, and ECM-remodeling enzymes.. During *Zebrafish* gastrulation, the coherent migration of prechordal plate progenitors relies on a fibrillar matrix comprised of FN and laminin. The

Fig. 3—Cont'd masculinization of the urethral tube. (B) **ECM porosity:** Spatiotemporal regulation of BM pore formation promotes cell migration and primitive streak extension in mice. During early stages of embryonic development, MMP-induced BM perforations are distributed throughout the BM overlying the two adjacent tissues – epiblast (pink) and extraembryonic ectoderm (blue). As development progresses, the expression of MMPs (blue) and the distribution of BM pores become asymmetrically localized to the posterior side of the embryo. During gastrulation, epiblast cells at the primitive streak undergo epithelial to mesenchymal transition, migrating through the BM pores to form mesoderm and endoderm. Blocking MMP activity inhibits BM pore formation, thereby impeding primitive streak extension. (C) **ECM dynamics:** A localized increase in BM turnover guides collective cell movement and *C. elegans* gonad elongation. During gonad morphogenesis, the leader cell, known as Distal Tip Cell (DTC) (green), secretes MMPs (blue) that degrade the matrix at the front of the gonadal tube. This allows a directional release of gonadal pressure and collective movement of proliferating germ cells (purple), present behind the DTC, within the confinement of the surrounding BM. Consequently, the gonad elongates along the anterior-posterior axis. (D) **Cell-matrix anchorage:** Integrin-based cell-matrix coupling is essential for efficient migration of the Zebrafish posterior lateral line (pLL) primordium (purple). The pLL primordium migrates under the skin (gray) and on top of the BM. Small, short-lived integrin clusters (green) form at the basal side of migrating cells. In integrin mutant embryos, primordium cells migrate at a reduced speed relative to wild-type embryos. This slower migration is primarily due to the decoupling of F-actin flow from the BM, disrupting effective force transduction and forward movement. Actin flow (red arrows) is significantly increased in primordium cells of integrin mutants. (E) **ECM stiffness:** Matrix stiffness influences the growth and directionality of *Xenopus laevis* retinal ganglion cell (RGC) axons. In developing *Xenopus* embryos, neurons (black lines) grow straight and parallel to each other in the stiff regions of the brain (yellow). They make a turn in the mid-diencephalon region, which has a gradient of stiffness. Eventually, they grow toward the softer optic tectum region, where the axons branch, splay apart, and establish synapses. Reducing stiffness by treating with chondroitin sulfate results in aberrant RGC growth and loss of directionality.

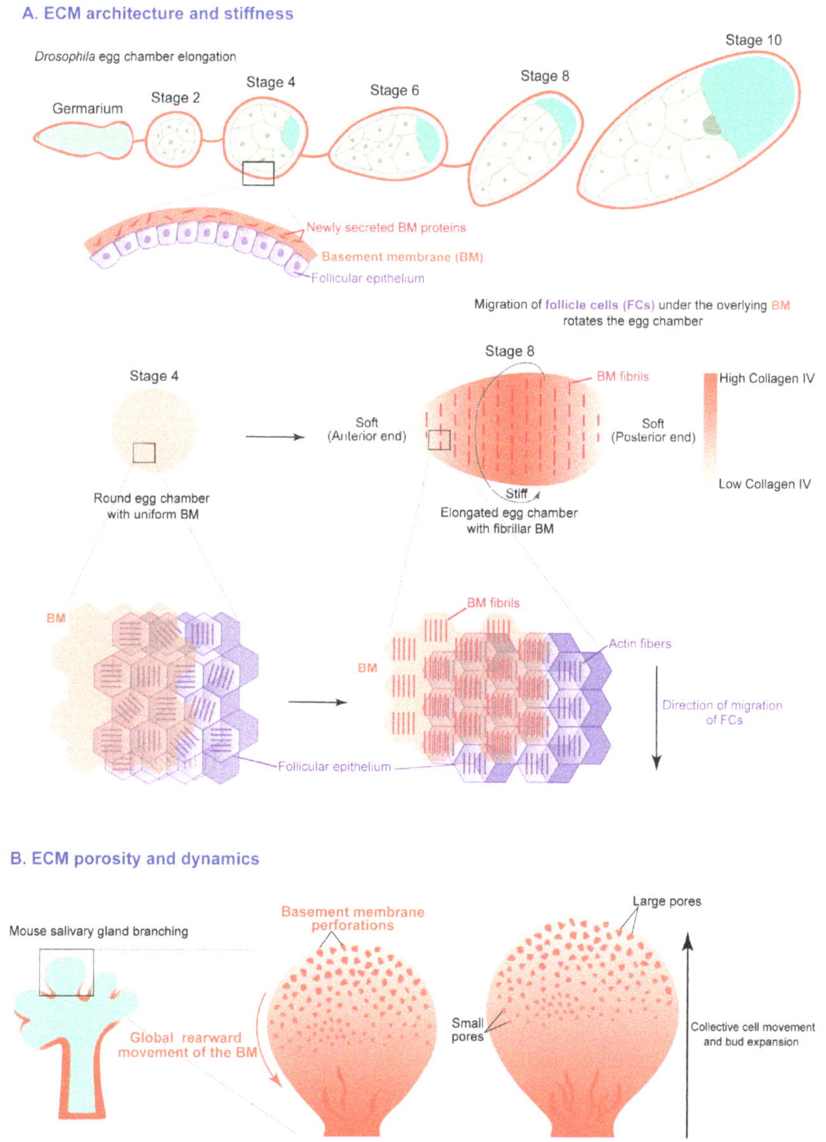

Fig. 4 Multiple ECM mechanical strategies collectively drive cell movement and tissue morphogenesis. (A) **ECM architecture and stiffness:** During *Drosophila* ovary development the egg chambers undergo a marked morphological transformation shifting from a spherical to an elongated shape. During elongation, collective migration of follicle cells within the surrounding BM rotates the egg chamber. The follicular epithelium secretes, and deposits BM fibrils aligned perpendicular to the anterior-posterior (AP) axis. Key drivers of egg chamber elongation include the reorganization of BM architecture from planar to fibrillar, an increase in BM stiffness

(Continued)

formation of this fibrillar pattern is regulated by another ECM protein known as EGF-containing fibulin-like extracellular matrix protein 2a (Efemp2)/Fibulin 4 (FBln4). In embryos deficient for Efemp2a, the fibrillar matrix collapses into disorganized clumps, leading to random and uncoordinated cell movement – a bit like trying to walk through a tangled forest without a clear trail. This disorganization results in developmental abnormalities, such as a misplaced head and a shortened body axis (Zhang et al., 2014).

A similar role in maintaining the ECM architecture has been proposed for a collagen receptor known as discoidin domain receptor 2 (DDR2) during craniofacial development in mice. DDR2 is expressed in the progenitor cells that give rise to cartilage-forming chondrocytes and bone-forming osteocytes during embryogenesis. DDR2-deficient mice exhibit flat skulls and reduced growth along the anterior-posterior axis. Immunostaining experiments and second harmonic generation imaging revealed that wild-type mice have cartilage matrix with homogenous fibrillar organization of Collagen II. In contrast, DDR2-deficient mice show misoriented and anisotropic collagen II fibrils. The disorganized collagen matrix leads to defects in chondrocyte polarization and results in craniofacial abnormalities (Mohamed et al., 2023).

MMPs also play a critical role in reshaping the ECM topology. During the development of external genitalia in mice, Mmp11, a matrix metalloprotease, degrades collagen VI. The reduction in collagen levels is responsible for a remodeling of FN from ring-like structures into a fibrillar matrix. Subsequently, FN fibrils promote mesenchymal cells to adopt a migratory behavior that is essential for the masculinization of the urethra (Fig. 3A) (Alcantara et al., 2022).

Fig. 4—Cont'd (resulting from higher Collagen IV levels), and a stiffness gradient (with higher Collagen IV concentration in the center that gradually decreases toward the poles). In later stages, BM fibrils serve as a guiding cue aligning the F-actin stress fibers of the follicle cells perpendicular to AP axis. Also, anterior follicle cells reorient along the AP axis, further facilitating egg chamber elongation. (B) **ECM porosity and dynamics**: During branching morphogenesis in mouse salivary glands, micro-perforations form in the BM surrounding the expanding epithelial buds. These perforations, created by MMPs, are concentrated at the bud tips. The size of BM pores decreases gradually from the tip to the duct region, where BM accumulates due to global rearward movement of the BM. This allows epithelial cell movement and bud expansion to remain restricted to the growing tips of the buds. Thus, both localized BM perforations and global BM dynamics contribute to collective cell movement and tissue growth.

How does fibrillar ECM topology promote cell migration? Aligned ECM fibers may guide uniaxial cell migration through the reorganization of the cytoskeleton. For instance, during *Drosophila* egg chamber elongation, BM fibrils act as a template that reorients F-actin stress fibers. The *Drosophila* egg chamber is composed of germ cells encased by an epithelium made of somatic follicle cells. This epithelium secretes a BM on its basal side, while its apical side is in contact with the germline. Live imaging has shown that follicle cells crawl over the underlying BM and rotate the egg chamber as it elongates along the anterior-posterior (AP) axis during the 14 stages of egg chamber development (Cetera, Lewellyn, & Horne-Badovinac, 2016). During stages 1–8, BM fibrils are deposited and aligned perpendicular to the AP axis of the egg chamber at the basal side of the follicle cells. An ECM receptor complex known as Dystroglycan (Dg)-Dystrophin (Dy) complex links the ECM to the F-actin cytoskeleton. This complex uses BM fibrils' orientation as a cue to guide the alignment of actin stress fibers within the follicle cells during the later stages of development (Fig. 4A) (Cerqueira Campos et al., 2020; Van De Bor et al., 2021; Zajac & Horne-Badovinac, 2022). The aligned stress fibers act as a "molecular corset", resisting circumferential expansion of germ cells, thereby forcing the egg chamber to elongate along the AP axis (Bilder & Haigo, 2012; Cetera & Horne-Badovinac, 2015; Cetera et al., 2014; Isabella & Horne-Badovinac, 2016). Whether aligned actin stress fibers also influence collective follicle cell migration remains ambiguous.

Clear evidence of orchestration of the cytoskeletal machinery by ECM alignment to guide cell migration has been obtained from in vitro studies. Using a simple two-dimensional (2D) fibrous ECM model, consisting of a collection of cell-adhesive fibronectin lines, Wang et al. discovered that ECM alignment influences various aspects of cell behavior. The extent of ECM alignment correlated with the elongation of cells, as well as increased speed, directionality, and persistence of migration. Moreover, cells exhibited highly organized focal adhesions and actin stress fibers, and active Rac1 localized preferentially along the direction of aligned ECM lines. Polarized Rac1 activity resulted in the formation of a large and stable protrusive front, leading to uniaxial directed cell movement (Wang et al., 2018).

Cells not only follow ECM cues, but also actively deform ECM topology to influence cell migration. In a deformable 2D synthetic matrix of aligned fibers, migrating fibroblasts used contractility to pull the fibers towards their cell body and by doing so stored elastic energy. Following matrix deformation, cell-matrix adhesions at the trailing edge were released, and the matrix recoiled, causing a sudden translocation of the cell body along the direction of the aligned fibers, analogous to a sling shot (Wang et al., 2019). In contrast, fibroblasts migrating within a 3D fibrous collagen gel exhibited an "anterior contraction," wherein pulling on collagen fibers was localized to the leading edge of the cell. This creates an anisotropically deformed or pre-strained matrix that promotes frontal protrusion, slow trailing end retraction, and migration in the direction of the ECM deformation. Disrupting this matrix strain at the front using localized laser incision impeded cell movement (Doyle et al., 2021). This mechanism differs from the classical model of mesenchymal cell migration mechanism, where rear-end retraction drives forward movement. Instead, in the 3D collagen matrix, fibroblast migration resembles a "front-wheel drive," driven by polarized frontal contraction and ECM deformation. Whether such dramatic mechanisms occur in living organisms is still an open question, but these in vitro findings give us a glimpse into how cells might harness the ECM architecture to travel more effectively.

4.2 ECM porosity

Porosity is a key physical characteristic of the ECM that refers to the extent of free spaces or gaps present within the matrix. Cells often move through pre-existing pores in the matrix. The secretion, degradation, alignment, and organization of ECM components all contribute to its overall porosity. Spatially restricted pores within the BM facilitate collective cell movement and expansion of branched mammalian organs such as lungs, salivary glands, and kidney (Harunaga et al., 2014; Wang et al., 2021). In the branching mouse salivary glands, BM micropores are found exclusively at the growing tips of the buds, while they are absent in the cleft and duct regions. The formation of these micropores depend on both local and global BM dynamics, which are regulated by metalloprotease-mediated matrix degradation and actomyosin contractility-driven matrix remodeling. Myosin II contractility likely physically pulls and remodels the BM. Also, photobleaching experiments revealed that the BM moves from the growing tips toward the duct. This movement results in a gradual decrease in pore size, which subsequently facilitates tissue extension at the tips of the developing buds and stabilizes the duct region (Fig. 4) (Harunaga et al., 2014).

Similarly, in early mouse embryos, BM perforations are formed in a spatiotemporally regulated manner to facilitate primitive streak extension (Kyprianou et al., 2020) (Fig. 3B). In the post-implantation embryo (E5–6.5), the epiblast and extraembryonic ectoderm – two adjacent tissues – are encased by a thin BM and visceral endoderm. Initially, BM pores are evenly distributed throughout the embryo, but by day E6.5, they become restricted to the posterior region. Formation of these pores depends on the expression of MMPs. Gastrulation starts at the most distal part of the posterior BM perforations, a region known as the primitive streak. Here, the primitive streak extends as epiblast cells undergo an epithelial-to-mesenchymal transition, migrate through the BM pores, and ingress between the epiblast and extra-embryonic ectoderm to form the mesoderm and endoderm. Inhibiting MMP activity disrupts BM pore formation, hinders primitive streak extension, and ultimately impairs the process of normal gastrulation (Fig. 3B) (Kyprianou et al., 2020) (Fig. 3B).

Having large ECM pores does not always mean faster or more efficient cell migration. A study by Soans et al. showed that increasing ECM pore size disrupts cell movement during zebrafish optic cup formation (Soans et al., 2022). During this process, rim cells migrate collectively over the underlying ECM using cryptic lamellipodia – protrusions that form at the basal side of the cell in the direction of migration. These structures are crucial for effective rim cell movement. Laminin staining revealed that the ECM beneath the migrating cells has a highly porous topology. Knocking down laminin or nidogen led to an increase in pore size. Interestingly, larger pores made it difficult for cryptic lamellipodia to form, causing the rim cells to get stuck in the matrix, lose their direction, and stall. This indicates that an optimal level of ECM porosity is needed for effective cell migration (Soans et al., 2022). It remains unclear how ECM porosity affects the formation of cell protrusions. Furthermore, it remains unexplored whether disrupting laminin or nidogen affects other physical features of the ECM that might also hinder cell movement.

Cells can utilize their cytoskeletal machinery to remodel and create wide gaps in the BM. This is observed when the anchor cell invades the vulval epithelium during *C. elegans* reproductive system development. The anchor cell is a uterine cell that creates an opening in the underlying BM to connect the uterine and vulval epithelium. While it normally uses proteases, the anchor cell can breach the BM even when protease activity is blocked. In the absence of matrix metalloproteases, it forms a large, actin-rich protrusion via the ARP2/3 complex that physically tears and breaches the BM (Cáceres et al., 2018; Kelley et al., 2019). Recent studies suggest that the anchor cell recruits mitochondria to the protrusive site to generate the energy required to power this forceful actin polymerization and breaching of the BM (Kelley et al., 2019). Similarly, human macrophages and cancer cells, cultured in vitro, were found to deploy actomyosin dependent forces to physically expand preformed pores, thus allowing their efficient migration (Bahr, Li, Feinberg, Jiang, & Weiss, 2022; Gong et al., 2021).

When matrix remodeling is not feasible, some cells adapt by altering their shape, or that of their nucleus, to squeeze through tight spaces. Since visualizing cell-matrix dynamics deep within tissues is technically challenging, most of the modes of migration discussed hereafter in this section have been observed in *vitro*. A perfect example of "shapeshifting" cells are leukocytes, which are proficient at rapidly changing their shape to probe and navigate through the ECM in paths of least resistance (Charras & Paluch, 2008; Renkawitz et al., 2019; Wolf et al., 2003). Under the conditions of high confinement and low cell-matrix adhesion, fibroblasts, cancer cells, and zebrafish gastrulating cells can quickly switch from a slow mesenchymal migration mode to a faster amoeboid mode (Bergert et al., 2015; Liu et al., 2015; Ruprecht et al., 2015). Mesenchymal cells typically have a large, protrusive front, while amoeboid cells are more rounded with unstable blebs at their front. This switch to amoeboid migration occurs due to increased cortical contractility, which generates a global retrograde flow of the cortex. Cortical flow coupled with the substrate drives stable bleb formation and forward cell movement through confined spaces (Liu et al., 2015; Ruprecht et al., 2015).

How do cells sense that they are in a tight place and increase their contractility in response? Two studies performed in cultured cells under confinement, revealed a mechanotransduction pathway that is initiated by tension in the nuclear envelope (Lomakin et al., 2020; Venturini et al., 2020). When cells are compressed beyond a critical threshold, their nuclear membrane unfolds, leading to the release of calcium ions (Ca^{2+}) from the endoplasmic reticulum into the cytoplasm. This rise in Ca^{2+} concentration activates nucleoplasmic phospholipase A2 (cPLA2), which helps generate arachidonic acid. Arachidonic acid, in turn, increases the contractility of the cell cortex, stiffens the cell and enables it to push through tight spaces (Lomakin et al., 2020; Venturini et al., 2020).

While the cytoplasm can stream through nearly any pore, the nucleus is a relatively large and stiff object and therefore deformation of the nucleus facilitates cell migration through dense ECM networks with a small pore size (Kalukula et al., 2022; Wolf et al., 2013). Nuclear deformation during 3D cell migration results from forces applied by the cytoskeletal machinery (McGregor, Hsia, & Lammerding, 2016). Increased actomyosin contractility at the rear end of the cell can propel the nucleus forward, allowing its movement through confined spaces (Lämmermann et al., 2008; Thomas et al., 2015). Alternatively, cells can generate force at the front, pulling the nucleus forward. This mechanism is often seen in lobopodial-based migration, where the nucleus acts like a piston. In this case, the pulling action relies on actomyosin bundles and intermediate filaments like vimentin, that link the nucleus to the cytoskeleton through the nuclear protein Nesprin (Davidson et al., 2020; Petrie, Koo, & Yamada, 2014; Petrie, Harlin, Korsak, & Yamada, 2017). This pulling force pressurizes the cell front, opening ion channels, which allow ions and water to flow in, further increasing the pressure. This leads to the expansion of the front protrusion, which wedges open the matrix for the cell to pass through (Lee et al., 2021). Additionally, cells may form a "perinuclear actin sleeve," created by the Arp2/3 complex. This sleeve compresses the nucleus from the sides, helping the cell move through narrow ECM pores (Thiam et al., 2016). Interestingly, metastatic cancer cells seem to have evolved a clever way to bypass the challenges of tight spaces. These cells have lower levels of the nucleoskeleton protein Lamin A/C, which normally stiffens the nuclear envelope. With less Lamin A/C, the nucleus becomes softer, allowing cancer cells to squeeze through constricted environments with greater ease, facilitating their invasion and spread (Bell et al., 2022; Harada et al., 2014).

4.3 ECM dynamics

In this section, ECM dynamics is defined in two ways. First, it is referred to in the terms of "turnover", which is the rate of synthesis and degradation of matrix. Second, it is defined as the "movement" of the matrix, which refers to the distance traveled by ECM components over time, either in a specific direction or randomly (Matsubayashi, 2022) (Fig. 2G).

In the previous section we discussed strategies used by cells migrating within dense ECM to squeeze through narrow holes. An alternative approach to alleviate the spatial confinement imposed on cells by the surrounding BM is to enhance the turnover rate of its constituent matrix components. This process may prevent the excessive accumulation of these components, thereby potentially reducing the overall BM density and allowing cells to navigate effectively. Recently, a study investigating *C. elegans* gonad morphogenesis, demonstrated that a localized increase in BM turnover mediated by MMPs secretion enables directed cell movement and tissue elongation (Agarwal, Shemesh, & Zaidel-Bar, 2022). The two U-shaped gonadal arms of *C. elegans* are enveloped by a BM surrounding a leader cell - known as the distal tip cell (DTC) - and follower proliferating germ cells. During larval development, proliferating germ cells exerts pressure on the leading DTC and the surrounding BM. The DTC secretes a metalloprotease called GON-1, which degrades the BM, specifically at its tip. As a result, the pressure within the gonad is released at the DTC front, leading to directional gonad elongation (Fig. 3C). Depletion of GON-1 causes abnormal accumulation of laminin, a BM component, that halts DTC movement and prevents gonad elongation. Consistently, photobleaching of fluorescently tagged laminin revealed that laminin is more mobile at the DTC front compared with the other regions of the gonad. Thus, enhanced BM dynamics at the DTC front allows its forward movement and gonad elongation (Agarwal et al., 2022).

Interestingly, FRAP experiments on the BM surrounding the pharynx of L4 stage *C. elegans* have revealed that different ECM components exhibit different dynamics. After photobleaching, fluorescently labeled endogenous collagen IV and laminin recovered slowly and uniformly across the bleached region, indicating that the fluorescence predominantly recovered from the extracellular fluid. In contrast, fibulin and nidogen recovered quickly, with recovery biased at the edge of the bleached region, suggesting that the fluorescent proteins moved from nearby unbleached region into the bleached region. Thus, it suggests a new model of BM architecture with dynamic BM proteins moving actively within a stable scaffold of the BM composed of collagen IV and laminin (Keeley et al., 2020). Does the dynamic diffusion of ECM components alter the rheological properties of the BM? What mechanisms regulate the differential dynamics of ECM components? These questions remain unanswered.

Besides ECM turnover, global movement of the BM also contributes to collective cell movement and tissue expansion, as observed during branching morphogenesis in mouse salivary glands (discussed in Section 4.2). A stable collagen IV network moves directionally from the top to the bottom of the growing salivary gland bud at an average speed of 8 μm/hr. This allows the cell movement and tissue growth to be localized to the growing tips of the salivary gland (Harunaga et al., 2014) (Fig. 4B). A similar global and directional movement of a fibrillin and fibronectin network has been observed during gastrulation in chick embryos (Czirók, Rongish, & Little, 2004; Zamir et al., 2006). How does ECM movement impact cell movement? Do they move together or relative to each other? To address these questions, Bortier et al. analyzed the movement of epiblast cells migrating towards the primitive streak in the gastrulating avian embryo (Bortier, Callebaut, Van Nueten, & Vakaet, 2001). The BM underlying the quail epiblast was labeled with radioactive glucosamine, and a fragment of this labeled epiblast was subsequently grafted into an unlabeled gastrulating chick embryo. If the epiblast and the BM move together, the radiolabel would remain confined to a region underneath the grafted cells. However, a trail of radioactive label was left behind as the epiblast cells moved towards the primitive streak. This suggests that the epiblast cells move relative to the matrix (Bortier et al., 2001). However, simultaneous imaging and tracking of GFP-expressing epiblast cells and antibody labeled fibronectin revealed that the ECM moves substantially during gastrulation, and the epiblast cells move in close concert with the moving matrix (Zamir et al., 2008).

4.4 Cell-matrix anchorage

One of the earliest theories explaining how adherent cells crawl on a two-dimensional surface, known as the 'molecular clutch' hypothesis, was introduced by Mitchison and Kirschner (Mitchison & Kirschner, 1988). They proposed that cells must adhere to the substrate to propel themselves forward. This adhesion is facilitated by cell-matrix adhesion receptors, such as integrins, which bind the ECM on the outside and can attach to the actin cytoskeleton on the inside, through adaptor proteins such as talin and vinculin. Actin polymerization at the leading edge creates a force that can push the membrane forward in large protrusions called lamellipodia, but only if the backward flow of actin is met by resistance. This counterforce is provided by the integrin adhesions anchored to the ECM when they engage with the actin and introduce friction. The 'molecular clutch' analogy refers to the regulated connection between integrins and the retrograde actin flow. When the clutch is engaged the actin polymerization force is harnessed for protrusion and migration, whereas when the clutch is disengaged the polymerizing actin flows rearward uninterrupted, the membrane remains in place, and the cell does not advance (Case & Waterman, 2015; Doyle et al., 2022) (Fig. 2H).

A recent study led by Holger Knaut and colleagues revealed the relevance of the molecular clutch model also for in vivo migration, at least when it takes place along BM. They showed that talin- and integrin-mediated coupling promote collective cell migration of the posterior lateral line primordium during *Zebrafish* development (Fig. 3D). The posterior lateral line primordium consists of a group of cells migrating in-between the skin and muscles along the anterior posterior axis and is stratified into two layers: superficial cells on top and basal cells underneath (Dalle Nogare & Chitnis, 2017). The basal cells are in close contact with the BM, which is composed of laminin and collagen IV. Live imaging revealed the presence of clusters of integrins and talin that colocalized with F-actin at the basal side of the primordium (Yamaguchi et al., 2022). When integrins were depleted, the actin network flowed backwards more rapidly, and cell movement became less efficient. Thus, linking the actin network to the matrix through integrin complexes provides the driving force for efficient primordium motility (Fig. 3D). Interestingly, in contrast to the larger and more persistent focal adhesions observed in cells migrating in vitro, the integrin clusters in the posterior lateral line primordium were notably smaller, measuring only 2 μm in size, with a lifetime of nearly 2 min (Yamaguchi et al., 2022). Reduced integrin activity has been shown to prevent the migration of macrophages to distant sites during *Drosophila* development (Comber et al., 2013; Siekhaus et al., 2010). Also in macrophages, actin flows in a polarized manner from the leading edge towards the cell body (Yolland et al., 2019) and colocalizes with the Collagen IV secreted by macrophages (Matsubayashi et al., 2017). Particle image velocimetry analysis indicated that migrating macrophages deform the surrounding ECM, suggesting they exert traction on it (Matsubayashi et al., 2017). However, without clear evidence of integrin localization in the macrophages, it remains uncertain whether the failure of these cells to spread in the embryo, upon integrin depletion, is due to lack of resistance to the actin flow by ECM-anchors.

During *C. elegans* gonad morphogenesis, integrin-mediated cell-matrix adhesions guide the directionality of the DTC without affecting its movement per se. As already mentioned above in Section 4.3, during larval stages, the DTC, a matrix-remodeling leader cell, is followed by proliferating germ cells that push it from behind. The DTC first migrates in a straight line parallel to the ventral surface of the worm body and then it makes a U-turn from the ventral towards the dorsal surface. This U-turn is caused by a polarized distribution of integrins, which concentrate on the dorsal side of the DTC. The anchoring of the DTC more strongly to the matrix on the dorsal side, with a concomitant push from the proliferating germ cells, generates torque that causes the DTC to rotate. Once the DTC completes a 180 degree turn, the integrins redistribute symmetrically and the gonad elongates in a straight line again. Disruption of this polarized distribution of integrins either prevents the turn or causes it to occur in an incorrect direction (Agarwal et al., 2022).

Integrin-mediated cell-matrix adhesion can, unexpectedly, act as a physical barrier to cell migration in a non-cell-autonomous manner. During early *Drosophila* embryo development, macrophages migrate between the ectoderm and mesoderm. A thin layer of laminin separates these two tissues, and cells from both sides attach to the BM through integrin adhesions. For macrophages to migrate effectively, these strong attachments to the ECM must be disrupted. Mitotically dividing ectodermal cells round up and disassemble integrin mediated cell-matrix adhesion adjacent to the mesodermal cells. This disassembly of cell-matrix adhesions creates a physical gap between the two adjacent tissues, enabling macrophages to squeeze through and invade the space (Akhmanova et al., 2022).

While the above-mentioned cases illustrate specific mechanisms by which integrin-mediated adhesions influence cell migration, integrins have also been shown to affect migration in various other developmental contexts although the underlying mechanisms are still not fully understood. During the development of mice enteric nervous system, a population of enteric neural crest cells (ENCCs) migrates from the dorsal side of the neural tube to the fore gut, caecum, and hindgut. After colonization, these ENCCs proliferate and differentiate to form neurons and glial cells. ß1 integrin depleted ENCCs stop migrating prematurely as they invade into the caecum and instead aggregate to form abnormal clusters (Breau, Dahmani, Broders-Bondon, Thiery, & Dufour, 2009). Similarly, integrins have been observed at the leading-edge protrusions of migrating cardiac cells during *Zebrafish* heart valve development (Gunawan et al., 2019). Endocardial cells move collectively into the ECM, and their cellular rearrangements eventually form valve leaflets. Depletion of integrin and talin in endocardial cells delays their migration and impairs valve morphogenesis (Gunawan et al., 2019). How exactly these cell-matrix links affect ENCCs and endocardial cell movement requires further research.

4.5 ECM Stiffness

ECM stiffness is a physical property defined by the ability of the matrix to resist deformation in response to an applied force. Matrix stiffness is determined by (1) ECM composition, (2) ECM deposition (3) connectivity of the network, resulting from activity of cross-linking enzymes and protease activity, (4) fiber organization and thickness. Each of these paramcters is tunable by the cells secreting and remodeling the ECM (Janmey et al., 2020).

A well-studied in vivo system, demonstrating the role of ECM stiffness in cell migration, is *Drosophila* oogenesis. As described previously, during egg chamber elongation (stage 1 to 14), follicle cells migrate collectively underneath a BM and rotate the egg chamber. During their rotational migration, they secrete BM fibrils aligned perpendicular to the AP axis. At the same time, there is an increase in laminin and collagen IV deposition. Atomic force microscopy studies revealed that increased deposition of BM components and BM fibril incorporation increases BM stiffness (Fig. 4A) (Chlasta et al., 2017; de la Loza et al., 2017). Manipulating BM stiffness by altering the levels of laminin affects the onset and migration speed of follicle cells (de la Loza et al., 2017). Atomic force microscopy has also revealed a stiffness gradient along the anterior-posterior axis, with the highest stiffness at the center of the egg chamber tapering toward the poles (Fig. 4A). This gradient contributes to egg chamber elongation (Crest, Diz-Muñoz, Chen, Fletcher, & Bilder, 2017). Interestingly, the softer BM at the anterior pole instructs collective cell migration and polarized reorientation of a subset of follicle cells from latitudinal towards parallel to the A-P axis, controlling global egg chamber elongation (Chen, Crest, Streichan, & Bilder, 2019). Genetic mutants with altered BM stiffness caused impaired cell reorientation and round egg chambers. In terms of signaling, mechanical cues from the BM are transduced to the follicle cells through Src tyrosine kinase signaling that alters junctional E-cadherin trafficking (Chen et al., 2019). How does cell reorientation, restricted to a subset of anterior follicle cells, contribute to organ elongation? And how do BM mechanics activate Src signaling? Answers to these questions necessitate further investigation.

The mechanical properties of the egg chamber BM also influence the migration of another group of follicle cells known as border cells (BCs). At around stage 9, when egg chamber rotation stalls, BCs delaminate from the anterior region of the follicle epithelium. They extend actin-rich protrusions and move collectively between a subset of germ cells called nurse cells until they contact the oocyte membrane. BC migration occurs in two phases: an initial streamlined and sliding phase, followed by a slower phase characterized by BC cluster tumbling and rotation. Manipulating the stiffness of the BM, by altering laminin or collagen IV levels surrounding the follicle cells, influences the cortical tension of both nurse and follicle cells. This in turn, alters the speed, dynamics and the mode of BC migration. In laminin-depleted egg chamber, BCs migrate faster toward the oocyte, display ectopic lateral protrusions, and approximately 40% of them fail to show rotational movement (Molina López et al., 2023). Thus, the mechanical properties of the ECM can affect the migration of cells that are not even in contact with it, through its effect on the mechanical properties of intermediary cells.

A key example of ECM stiffness guiding cell migration during vertebrate development comes from studies in *Xenopus* embryos (Koser et al., 2016). In the developing brain of *Xenopus* embryos, retinal ganglion cell axons navigate along a stereotypical pathway. They exit the retina via the optic nerve, travel along a region of optic tract (OT), make a caudal turn in the mid-diencephalon region, and finally terminate at the optic tectum (Fig. 3E). Using an AFM-based approach, Koser et al. mapped the stiffness of the brain tissue. They found a gradient of stiffness along the axon trajectory with the OT being stiffer than the tectum. Axons grow towards the softer tectum and are mechanosensitive. Manipulating chondroitin sulfate proteoglycans, an ECM component, reduced overall tissue stiffness and disrupted retinal axon growth and directionality (Koser et al., 2016) (Fig. 3E). It is important to note that, although the ECM contributes to tissue stiffness, the stiffness gradient observed here is not due to the ECM. Instead, it results from differences in cell density. DAPI staining revealed that the stiffer OT region has a significantly higher cell density than the tectum. Inhibiting mitosis disrupted this stiffness gradient and interfered with axon turning (Thompson et al., 2019). Another study revealed the presence of a stiffness gradient in the early mouse limb bud. During early limb morphogenesis, mesodermal cells migrate from the lateral plate region towards the limb field to form the characteristic limb bud shape. The collective movement of mesodermal cells was biased towards the increased stiffness, corresponding to a region of high expression of fibronectin (Zhu et al., 2020). However, it has not yet been tested whether elevated fibronectin levels directly contribute to increased tissue stiffness, and if altering the stiffness gradient affects limb cell migration. Together, these studies emphasize the essential role of ECM stiffness in guiding cell migration during development but whether stiffness gradients of the ECM itself, and not of the tissue, affect cellular migration dynamics in vivo is unknown.

How do cells sense and respond to stiffness? The ability of cells to migrate towards a stiffer substrate is defined as "durotaxis" (Lo et al., 2000). Due to the relative ease of manipulation of matrix rigidity in vitro, durotaxis is a well-recognized phenomenon in multiple cell types (Espina et al., 2022; Janmey et al., 2020). One of the classic explanations for durotaxis is based on the "molecular clutch" model (as explained earlier). On a stiffer matrix, the pulling force generated by actomyosin contractility strengthens the binding of integrins to the matrix. This leads to recruitment of more integrins and focal adhesion proteins like talin and vinculin which, in turn, results in increased traction force generation and forward cell movement. Conversely, on a softer matrix, fewer integrin-binding proteins are recruited, and there is an increase in focal adhesion disassembly. This establishes a biased movement of the cell towards stiffer matrix. Recently, several other molecular and physical models have been proposed to explain durotaxis (Espina et al., 2022; Ji & Huang, 2023; Shellard & Mayor, 2021; Sunyer & Trepat, 2020). Given that synthetic systems do not fully recapitulate the complexity of 3D tissue microenvironments, future challenges lie in validating these models or uncovering novel mechanisms that cells employ to sense ECM stiffness in vivo.

5. Outlook

Traditionally, the ECM has been assumed to provide structural support and/or act as a repository of biochemical cues that steer migrating cells (Hynes, 2009; Yamada et al., 2019). This review highlights the mechanical role of the ECM in modulating cell migration. Despite these advances, deciphering how the mechanical properties of the matrix influence cell behavior in native and dynamic 3D tissue microenviroment remains a challenge.

One of the major struggles in elucidating the role of ECM stems from its complexity and diversity across tissues and organs. Omics approaches (genomic sequencing and proteomics) combined with in silico bioinfomatic analysis have identifed the "matrisome" – a comprehensive collection of structural components of ECM such as collagen, proteoglycans, glycoproteins, etc., and ECM-associated proteins inlcuding proteases, matrix metalloproteases, cross-linking enzymes, growth factors, and others – across species (Hynes & Naba, 2012; Jayadev et al., 2022; Naba et al., 2012; Naba, 2023; Petrov, Considine, Izzi, & Naba, 2023; Shao et al., 2023) (https://matrisome.org). This project has discovered nearly 1000 matrisome genes in humans, mice, and *Zebrafish*. The intricacy of the ECM is not solely due to its various core matrisome components but is further amplified by the presence of multiple ECM protein isoforms. These isoforms arise from processes like alternative splicing, post-translational modifications, and enzymatic remodeling carried out by diverse ECM modulators (some of them are discussed in Section 3)(Hynes, 2009; Rekad, Izzi, Lamba, Ciais, & Van Obberghen-Schilling, 2022; Saraswathibhatla et al., 2023).

Another difficulty arises due to complex protein-protein interactions and their inter-dependant roles. Most studies in the literature have focused on manipulating individual ECM components to explore their roles in cell migration. However, recent research has shown that disrupting one ECM protein can affect several other ECM components, altering the entire matrix's composition, organization, and possibly mechanical properties of the ECM (Alcantara et al., 2022; Keeley et al., 2020; Zhang et al., 2014). As demonstrated by multiple examples in Section 4, manipulation of MMPs can alter various characteristics of the ECM, such as its porosity, turnover, density, and organization. Hence, cautious interpretations of matrix mechanics is required following the depletion of ECM components.

Telling apart the biochemical and mechanical roles of the ECM in vivo is a major concern in the field. ECM provides specific binding sites for morphogens, such as epidermal growth factors, Wnt proteins and fibroblast growth factors. Controlled release of these signaling molecules from the matrix creates a morphogen gradient that guides several cellular processes and tissue morphogenesis (Entchev & González-Gaitán, 2002; Sherwood, 2021). Thus, ECM modulators, such as cross-linking enzymes and proteases, can alter both its physical and signaling roles. As discussed earlier, MMPs facilitate collective cell movement either by creating gaps or by enhancing BM turnover (Agarwal et al., 2022; Harunaga et al., 2014; Kyprianou et al., 2020). Additionally, MMPs may release ECM-bound fragments or growth factors, influencing cell migration (Walker et al., 2018). For example, the degradation of collagen releases fragments like tumstatin and endostatin that affect cell movement, proliferation, and survival (Ortega & Werb, 2002). Transforming growth factor-β (TGF-β) is stored in an inactive form within the ECM. MMP-mediated degradation of matrix releases active TGF-β. This active TGF-β promote cancer cell migration and metastasis (Imai et al., 1997; Kessenbrock et al., 2015). Also, cleaved fragments of laminin bind to epidermal growth factor receptor to enhance cell migration during wound healing and cancer metastasis (Giannelli et al., 1997; Panayotou, End, Aumailley, Timpl, & Engel, 1989; Tran et al., 2005). Thus, whether the ECM provides a permissive physical microenvironment facilitating cell movement or if it plays an instructive role needs to be carefully examined in each system of interest.

Limited avaiblilty of tools and approaches to precisely measure and manipulate the mechanical properties of the ECM in vivo presents another hurdle in studying their role in cell migration. While AFM is predominantly used to measure stiffness *in* vivo, it cannot distinguish whether the measured stiffness arises from the ECM or the underlying cells. Another limitation is that the AFM measurements are taken orthogonally to the tissue plane, whereas migrating cells are likely to perceive stiffness within the plane of the tissue. Other techniques used for the biophysical characterization of developing tissues include magnetic beads, ferrofluid microdroplets, and optical tweezers (Bambardekar, Clément, Blanc, Chardès, & Lenne, 2015; Doubrovinski et al., 2017; Serwane et al., 2017). However, they have not been widely adopted due to technical challenges. Brillouin microscopy, a recently developed non-invasive and label-free approach, presents a promising alternative for measuring stiffness in vivo (Prevedel, Diz-Muñoz, Ruocco, & Antonacci, 2019). This optical technique leverages the interaction between incident light and acoustic phonons – thermally excited sound waves within a material – to probe the mechanical properties of the tissue. As light passes through a sample, some of it is scattered inelastically by these phonons, resulting in a frequency shift known as the Brillouin shift. This shift provides information on the tissue's viscoelastic properties, such as its stiffness and elasticity. Brillouin microscopy enables 3D imaging of the material properties of the tissues at high spatial and temporal resolution during development (Bevilacqua et al., 2021, 2023; Prevedel et al., 2019).

Despite the aforementioned technical limitations, recent advances using the CRISPR-Cas9 approach to label endogenous ECM proteins, combined with advanced live imaging, have provided novel insights into ECM dynamics and structure. Two studies published in 2020 (Keeley et al., 2020; Matsubayashi et al., 2020) revealed that ECM proteins are highly dynamic, with a much faster turnover rate (minutes to hours) compared to previous reports (months to years)(Decaris et al., 2014; Trier et al., 1990). Comprehensive labeling of 17 BM components also revealed that some of the BM components move actively while others are more static and stable constituents of the BM (Keeley et al., 2020) (for details see Section 4.3). However, the significance of ECM dynamics in cell migration still needs to be clarified.

Furthermore, mathematical and computational models of the ECM have been succesfully employed to understand how ECM mechanics influence cell migration, tissue structure and morphogenesis (Crossley et al., 2024; Díaz-de-la-Loza & Stramer, 2024). These models offer a complementary means of investigation to experimental approaches. They adddress the complexicity of ECM by selectively focussing on one or more features of the ECM and thereby, assisting to dissect its role in cell migration. For instance, Martinson et al. used an agent-based model to explore the role of ECM remodeling in collective NCC migration (also discussed in Section 4.1). Their simulation modeled NCCs as agents migrating on a 2D surface and incorporated features such as ECM remodeling, contact guidance, cell-cell repulsion, and haptotaxis. The model predicted that leader cells remodel punctate fibronectin into fibrillar FN matrix, facilitating long-range collective migration without causing cell jamming (Martinson et al., 2023). *In* silico models provide an efficient means to generate and test multiple hypotheses with minimal experimental work, ultimately saving time and resources. However, better tools and techniques are needed to characterize the mechanical properties of ECM in vivo in order to enable the design of more accurate and predictive models.

Moving forward, a comprehensive understanding of the mechanical role of the ECM in cell migration during development will require a multidisciplinary approach. This approach must integrate simultaneous visualization of ECM composition and dynamics, precise measurement of its biophysical properties, targeted genetic manipulation, and advanced computational modeling. Cell migration is not only fundamental to the formation of tissues and organs with diverse shapes and functions, but it also plays a pivotal role in pathological conditions such as cancer metastasis, immune disorders, and neurological diseases. Therefore, exploring the impact of ECM mechanics on cell migration will not only deepen our understanding of its role in morphogenesis but will also open new avenues for the development of therapeutic strategies to combat human diseases.

References

Agarwal, P., Shemesh, T., & Zaidel-Bar, R. (2022). Directed cell invasion and asymmetric adhesion drive tissue elongation and turning in C. elegans gonad morphogenesis. *Developmental Cell, 57*, 2111–2126.e6. https://doi.org/10.1016/j.devcel.2022.08.003.

Agarwal, P., & Zaidel-Bar, R. (2019). Diverse roles of non-muscle myosin II contractility in 3D cell migration. *Essays in Biochemistry, 63*, 497–508. https://doi.org/10.1042/EBC20190026.

Akhmanova, M., Emtenani, S., Krueger, D., Gyoergy, A., Guarda, M., Vlasov, M., ... De Renzis, S. (2022). Cell division in tissues enables macrophage infiltration. *Science (New York, N. Y.), 376*, 394–396.

Alcantara, M. C., Suzuki, K., Acebedo, A. R., Kajioka, D., Hirohata, S., Kaisho, T., ... Yamada, G. (2022). Androgen-regulated MafB drives cell migration via MMP11-dependent extracellular matrix remodeling in mice. *Iscience, 25*.

Apte, S. S. (2009). A disintegrin-like and metalloprotease (reprolysin-type) with thrombospondin type 1 motif (ADAMTS) superfamily: Functions and mechanisms. *Journal of Biological Chemistry, 284*, 31493–31497. https://doi.org/10.1074/jbc.R109.052340.

Aumailley, M., Wiedemann, H., Mann, K., & Timpl, R. (1989). Binding of nidogen and the laminin-nidogen complex to basement membrane collagen type IV. *European Journal of Biochemistry, 184*, 241–248. https://doi.org/10.1111/j.1432-1033.1989.tb15013.x.

Bahr, J. C., Li, X.-Y., Feinberg, T. Y., Jiang, L., & Weiss, S. J. (2022). Divergent regulation of basement membrane trafficking by human macrophages and cancer cells. *Nature Communications, 13*, 6409.

Bambardekar, K., Clément, R., Blanc, O., Chardès, C., & Lenne, P.-F. (2015). Direct laser manipulation reveals the mechanics of cell contacts in vivo. *Proceedings of the National Academy of Sciences of the United States of America, 112*, 1416–1421. https://doi.org/10.1073/pnas.1418732112.

Ban, E., Franklin, J. M., Nam, S., Smith, L. R., Wang, H., Wells, R. G., ... Shenoy, V. B. (2018). Mechanisms of plastic deformation in collagen networks induced by cellular forces. *Biophysical Journal, 114*, 450–461. https://doi.org/10.1016/j.bpj.2017.11.3739.

Baneyx, G., Baugh, L., & Vogel, V. (2002). Fibronectin extension and unfolding within cell matrix fibrils controlled by cytoskeletal tension. *Proceedings of the National Academy of Sciences of the United States of America, 99*, 5139–5143. https://doi.org/10.1073/pnas.072650799.

Barker, H. E., Cox, T. R., & Erler, J. T. (2012). The rationale for targeting the LOX family in cancer. *Nature Reviews. Cancer, 12*, 540–552. https://doi.org/10.1038/nrc3319.

Barsigian, C., Fellin, F. M., Jain, A., & Martinez, J. (1988). Dissociation of fibrinogen and fibronectin binding from transglutaminase-mediated cross-linking at the hepatocyte surface. *Journal of Biological Chemistry, 263*, 14015–14022. https://doi.org/10.1016/S0021-9258(18)68178-5.

Bell, E. S., Shah, P., Zuela-Sopilniak, N., Kim, D., Varlet, A.-A., Morival, J. L., ... Elacqua, J. J. (2022). Low lamin A levels enhance confined cell migration and metastatic capacity in breast cancer. *Oncogene, 41*, 4211–4230.

Bergert, M., Erzberger, A., Desai, R. A., Aspalter, I. M., Oates, A. C., Charras, G., ... Paluch, E. K. (2015). Force transmission during adhesion-independent migration. *Nature Cell Biology*. https://doi.org/10.1038/ncb3134.

Bevilacqua, C., Gomez, J. M., Fiuza, U.-M., Chan, C. J., Wang, L., Hambura, S., ... Prevedel, R. (2023). High-resolution line-scan Brillouin microscopy for live imaging of mechanical properties during embryo development. *Nature Methods, 20*, 755–760. https://doi.org/10.1038/s41592-023-01822-1.

Bevilacqua, C., Hambura, S., Wang, L., Chan, C. J., Eguren, M., Gomez Elliff, J. M., & Prevedel, R. (2021). High-resolution line-scanning Brillouin microscopy for fast and low phototoxicity live-imaging of mechanical properties in biology. In Larin, K. V., Scarcelli, G. (Eds.), *Optical Elastography and Tissue Biomechanics VIII*. Presented at the Optical Elastography and Tissue Biomechanics VIII, SPIE, Online Only, United States, p. 8. https://doi.org/10.1117/12.2582722.

Bilder, D., & Haigo, S. L. (2012). Expanding the morphogenetic repertoire: Perspectives from the Drosophila egg. *Developmental Cell, 22*, 12–23. https://doi.org/10.1016/j.devcel.2011.12.003.

Bonnans, C., Chou, J., & Werb, Z. (2014). Remodelling the extracellular matrix in development and disease. *Nature Reviews. Molecular Cell Biology, 15*, 786–801. https://doi.org/10.1038/nrm3904.

Bortier, H., Callebaut, M., Van Nueten, E., & Vakaet, L. (2001). Autoradiographic evidence for the sliding of the upper layer over the basement membrane in chicken blastoderms during gastrulation. *European Journal of Morphology, 39*, 91–98. https://doi.org/10.1076/ejom.39.2.91.7369.

Bradshaw, M. J., Hoffmann, G. A., Wong, J. Y., & Smith, M. L. (2019). Fibronectin fiber creep under constant force loading. *Acta Biomaterialia, 88*, 78–85. https://doi.org/10.1016/j.actbio.2019.02.022.

Breau, M. A., Dahmani, A., Broders-Bondon, F., Thiery, J. -P., & Dufour, S. (2009). β1 integrins are required for the invasion of the caecum and proximal hindgut by enteric neural crest cells.

Brown, N. H. (2011). Extracellular matrix in development: Insights from mechanisms conserved between invertebrates and vertebrates. *Cold Spring Harbor Perspectives in Biology, 3*, a005082. https://doi.org/10.1101/cshperspect.a005082.

Cáceres, R., Bojanala, N., Kelley, L. C., Dreier, J., Manzi, J., Di Federico, F., ... Sherwood, D. R. (2018). Forces drive basement membrane invasion in Caenorhabditis elegans. *Proceedings of the National Academy of Sciences, 115*, 11537–11542.

Cardoso, I., Stamnaes, J., Andersen, J. T., Melino, G., Iversen, R., & Sollid, L. M. (2015). Transglutaminase 2 interactions with extracellular matrix proteins as probed with celiac disease autoantibodies. *The FEBS Journal, 282*, 2063–2075. https://doi.org/10.1111/febs.13276.

Case, L. B., & Waterman, C. M. (2015). Integration of actin dynamics and cell adhesion by a three-dimensional, mechanosensitive molecular clutch. *Nature Cell Biology, 17*, 955–963.

Cerqueira Campos, F., Dennis, C., Alégot, H., Fritsch, C., Isabella, A., Pouchin, P., ... Mirouse, V. (2020). Oriented basement membrane fibrils provide a memory for F-actin planar polarization via the Dystrophin-Dystroglycan complex during tissue elongation. *Development (Cambridge, England), 147*, dev186957.

Cetera, M., & Horne-Badovinac, S. (2015). Round and round gets you somewhere: collective cell migration and planar polarity in elongating Drosophila egg chambers. *Current Opinion in Genetics & Development, 32*, 10–15. https://doi.org/10.1016/j.gde.2015.01.003.

Cetera, M., Lewellyn, L., & Horne-Badovinac, S. (2016). Cultivation and live imaging of Drosophila ovaries. In C. Dahmann (Ed.). *Drosophila, methods in molecular biology* (pp. 215–226). New York, NY: Springer New York. https://doi.org/10.1007/978-1-4939-6371-3_12.

Cetera, M., Ramirez-San Juan, G. R., Oakes, P. W., Lewellyn, L., Fairchild, M. J., Tanentzapf, G., ... Horne-Badovinac, S. (2014). Epithelial rotation promotes the global alignment of contractile actin bundles during Drosophila egg chamber elongation. *Nature Communications, 5*, 5511.

Charras, G., & Paluch, E. (2008). Blebs lead the way: How to migrate without lamellipodia. *Nature Reviews. Molecular Cell Biology*. https://doi.org/10.1038/nrm2453.

Chaudhuri, O., Cooper-White, J., Janmey, P. A., Mooney, D. J., & Shenoy, V. B. (2020). Effects of extracellular matrix viscoelasticity on cellular behaviour. *Nature, 584*, 535–546.

Chen, D.-Y., Crest, J., Streichan, S. J., & Bilder, D. (2019). Extracellular matrix stiffness cues junctional remodeling for 3D tissue elongation. *Nature Communications, 10*, 3339. https://doi.org/10.1038/s41467-019-10874-x.

Chlasta, J., Milani, P., Runel, G., Duteyrat, J.-L., Arias, L., Lamiré, L.-A., ... Grammont, M. (2017). Variations in basement membrane mechanics are linked to epithelial morphogenesis. *Development (Cambridge, England)*, dev.152652. https://doi.org/10.1242/dev.152652.

Comber, K., Huelsmann, S., Evans, I., Sánchez-Sánchez, B. J., Chalmers, A., Reuter, R., ... Martín-Bermudo, M. D. (2013). A dual role for the βPS integrin myospheroid in mediating Drosophila embryonic macrophage migration. *Journal of Cell Science, 126*, 3475–3484.

Crest, J., Diz-Muñoz, A., Chen, D.-Y., Fletcher, D. A., & Bilder, D. (2017). Organ sculpting by patterned extracellular matrix stiffness. *eLife, 6*. https://doi.org/10.7554/eLife.24958.

Crossley, R. M., Johnson, S., Tsingos, E., Bell, Z., Berardi, M., Botticelli, M., ... Shuttleworth, R. (2024). Modeling the extracellular matrix in cell migration and morphogenesis: A guide for the curious biologist. *Frontiers in Cell and Developmental Biology, 12*, 1354132. https://doi.org/10.3389/fcell.2024.1354132.

Czirók, A., Rongish, B. J., & Little, C. D. (2004). Extracellular matrix dynamics during vertebrate axis formation. *Developmental Biology, 268*, 111–122. https://doi.org/10.1016/j.ydbio.2003.09.040.

Dalle Nogare, D., & Chitnis, A. B. (2017). A framework for understanding morphogenesis and migration of the zebrafish posterior Lateral Line primordium. *Mechanisms of Development, 148*, 69–78.

Dalton, C. J., & Lemmon, C. A. (2021). Fibronectin: Molecular structure, fibrillar structure and mechanochemical signaling. *Cells, 10*, 2443. https://doi.org/10.3390/cells10092443.

Davidson, P. M., Battistella, A., Déjardin, T., Betz, T., Plastino, J., Borghi, N., ... Sykes, C. (2020). Nesprin-2 accumulates at the front of the nucleus during confined cell migration. *EMBO Reports, 21*, e49910.

de la Loza, M. C. D., Díaz-Torres, A., Zurita, F., Rosales-Nieves, A. E., Moeendarbary, E., Franze, K., ... Gonzalez-Reyes, A. (2017). Laminin levels regulate tissue migration and anterior-posterior polarity during egg morphogenesis in Drosophila. *Cell Reports, 20*, 211–223.

Decaris, M. L., Gatmaitan, M., FlorCruz, S., Luo, F., Li, K., Holmes, W. E., ... Emson, C. L. (2014). Proteomic analysis of altered extracellular matrix turnover in bleomycin-induced pulmonary fibrosis. *Molecular & Cellular Proteomics, 13*, 1741–1752. https://doi.org/10.1074/mcp.M113.037267.

Díaz-de-la-Loza, M.-C., & Stramer, B. M. (2024). The extracellular matrix in tissue morphogenesis: No longer a backseat driver. *Cells & Development, 177*, 203883. https://doi.org/10.1016/j.cdev.2023.203883.

Doubrovinski, K., Swan, M., Polyakov, O., & Wieschaus, E. F. (2017). Measurement of cortical elasticity in *Drosophila melanogaster* embryos using ferrofluids. *Proceedings of the National Academy of Sciences of the United States of America, 114*, 1051–1056. https://doi.org/10.1073/pnas.1616659114.

Doyle, A. D., Nazari, S. S., & Yamada, K. M. (2022). Cell–extracellular matrix dynamics. *Physical Biology, 19*, 021002. https://doi.org/10.1088/1478-3975/ac4390.

Doyle, A. D., Sykora, D. J., Pacheco, G. G., Kutys, M. L., & Yamada, K. M. (2021). 3D mesenchymal cell migration is driven by anterior cellular contraction that generates an extracellular matrix prestrain. *Developmental Cell, 56*, 826–841.

Dzobo, K., & Dandara, C. (2023). The extracellular matrix: Its composition, function, remodeling, and role in tumorigenesis. *Biomimetics, 8*, 146. https://doi.org/10.3390/biomimetics8020146.

Elosegui-Artola, A. (2021). The extracellular matrix viscoelasticity as a regulator of cell and tissue dynamics. *Current Opinion in Cell Biology, 72*, 10–18. https://doi.org/10.1016/j.ceb.2021.04.002.

Entchev, E. V., & González-Gaitán, M. A. (2002). Morphogen gradient formation and vesicular trafficking. *Traffic (Copenhagen, Denmark), 3*, 98–109. https://doi.org/10.1034/j.1600-0854.2002.030203.x.

Espina, J. A., Marchant, C. L., & Barriga, E. H. (2022). Durotaxis: The mechanical control of directed cell migration. *The FEBS Journal, 289*, 2736–2754. https://doi.org/10.1111/febs.15862.

Espitia Pinzón, N., Brevé, J. J. P., Bol, J. G. J. M., Drukarch, B., Baron, W., & Van Dam, A.-M. (2017). Tissue transglutaminase in astrocytes is enhanced by inflammatory mediators and is involved in the formation of fibronectin fibril-like structures. *Journal of Neuroinflammation, 14*, 260. https://doi.org/10.1186/s12974-017-1031-2.

Fidler, A. L., Boudko, S. P., Rokas, A., & Hudson, B. G. (2018). The triple helix of collagens – An ancient protein structure that enabled animal multicellularity and tissue evolution. *Journal of Cell Science, 131*, jcs203950. https://doi.org/10.1242/jcs.203950.

Filipe, E. C., Chitty, J. L., & Cox, T. R. (2018). Charting the unexplored extracellular matrix in cancer. *International Journal of Experimental Pathology, 99*, 58–76. https://doi.org/10.1111/iep.12269.

Fox, J. W., Mayer, U., Nischt, R., Aumailley, M., Reinhardt, D., Wiedemann, H., ... Engel, J. (1991). Recombinant nidogen consists of three globular domains and mediates binding of laminin to collagen type IV. *The EMBO Journal, 10*, 3137–3146. https://doi.org/10.1002/j.1460-2075.1991.tb04875.x.

Frantz, C., Stewart, K. M., & Weaver, V. M. (2010). The extracellular matrix at a glance. *Journal of Cell Science, 123*, 4195–4200.

Geiger, B., & Yamada, K. M. (2011). Molecular architecture and function of matrix adhesions. *Cold Spring Harbor Perspectives in Biology, 3*, a005033. https://doi.org/10.1101/cshperspect.a005033.

Giannelli, G., Falk-Marzillier, J., Schiraldi, O., Stetler-Stevenson, W. G., & Quaranta, V. (1997). Induction of cell migration by matrix metalloprotease-2 cleavage of laminin-5. *Science (New York, N. Y.), 277*, 225–228. https://doi.org/10.1126/science.277.5323.225.

Gong, Z., Wisdom, K. M., McEvoy, E., Chang, J., Adebowale, K., Price, C. C., ... Shenoy, V. B. (2021). Recursive feedback between matrix dissipation and chemo-mechanical signaling drives oscillatory growth of cancer cell invadopodia. *Cell Reports, 35*, 109047. https://doi.org/10.1016/j.celrep.2021.109047.

Gunawan, F., Gentile, A., Fukuda, R., Tsedeke, A. T., Jiménez-Amilburu, V., Ramadass, R., ... Stainier, D. Y. (2019). Focal adhesions are essential to drive zebrafish heart valve morphogenesis. *Journal of Cell Biology, 218*, 1039–1054.

Harada, T., Swift, J., Irianto, J., Shin, J.-W., Spinler, K. R., Athirasala, A., ... Discher, D. E. (2014). Nuclear lamin stiffness is a barrier to 3D migration, but softness can limit survival. *Journal of Cell Biology, 204*, 669–682.

Harunaga, J. S., Doyle, A. D., & Yamada, K. M. (2014). Local and global dynamics of the basement membrane during branching morphogenesis require protease activity and actomyosin contractility. *Developmental Biology, 394*, 197–205. https://doi.org/10.1016/j.ydbio.2014.08.014.

Hopf, M., Göhring, W., Mann, K., & Timpl, R. (2001). Mapping of binding sites for nidogens, fibulin-2, fibronectin and heparin to different IG modules of perlecan 1. Edited by M. F. Moody. *Journal of Molecular Biology, 311*, 529–541. https://doi.org/10.1006/jmbi.2001.4878.

Huet-Calderwood, C., Rivera-Molina, F. E., Toomre, D. K., & Calderwood, D. A. (2023). Fibroblasts secrete fibronectin under lamellipodia in a microtubule- and myosin II–dependent fashion. *Journal of Cell Biology, 222*, e202204100. https://doi.org/10.1083/jcb.202204100.

Humphries, J. D., Byron, A., & Humphries, M. J. (2006). Integrin ligands at a glance. *Journal of Cell Science, 119*, 3901–3903. https://doi.org/10.1242/jcs.03098.

Hynes, R. O. (2009). The extracellular matrix: Not just pretty fibrils. *Science (New York, N. Y.), 326*, 1216–1219. https://doi.org/10.1126/science.1176009.

Hynes, R. O., & Naba, A. (2012). Overview of the matrisome—An inventory of extracellular matrix constituents and functions. *Cold Spring Harbor Perspectives in Biology, 4*, a004903. https://doi.org/10.1101/cshperspect.a004903.

Imai, K., Hiramatsu, A., Fukushima, D., Pierschbacher, M. D., & Okada, Y. (1997). Degradation of decorin by matrix metalloproteinases: Identification of the cleavage sites, kinetic analyses and transforming growth factor- β 1 release. *Biochemical Journal, 322*, 809–814. https://doi.org/10.1042/bj3220809.

Isabella, A. J., & Horne-Badovinac, S. (2016). Rab10-mediated secretion synergizes with tissue movement to build a polarized basement membrane architecture for organ morphogenesis. *Developmental Cell, 38*, 47–60. https://doi.org/10.1016/j.devcel.2016.06.009.

Janmey, P. A., Fletcher, D. A., & Reinhart-King, C. A. (2020). Stiffness sensing by cells. *Physiological Reviews, 100*, 695–724. https://doi.org/10.1152/physrev.00013.2019.

Jayadev, R., Morais, M. R. P. T., Ellingford, J. M., Srinivasan, S., Naylor, R. W., Lawless, C., ... Lennon, R. (2022). A basement membrane discovery pipeline uncovers network complexity, regulators, and human disease associations. *Science Advances, 8*. https://doi.org/10.1126/sciadv.abn2265.

Jayadev, R., & Sherwood, D. R. (2017). Basement membranes. *Current Biology, 27*, R207–R211. https://doi.org/10.1016/J.CUB.2017.02.006.

Ji, C., & Huang, Y. (2023). Durotaxis and negative durotaxis: Where should cells go? *Communications Biology, 6*, 1169. https://doi.org/10.1038/s42003-023-05554-y.

Kalukula, Y., Stephens, A. D., Lammerding, J., & Gabriele, S. (2022). Mechanics and functional consequences of nuclear deformations. *Nature Reviews. Molecular Cell Biology, 23*, 583–602.

Keeley, D. P., Hastie, E., Jayadev, R., Kelley, L. C., Chi, Q., Payne, S. G., ... Sherwood, D. R. (2020). Comprehensive endogenous tagging of basement membrane components reveals dynamic movement within the matrix scaffolding. *Developmental Cell, 54*, 60–74. https://doi.org/10.1016/j.devcel.2020.05.022.

Kelley, L. C., Chi, Q., Cáceres, R., Hastie, E., Schindler, A. J., Jiang, Y., ... Sherwood, D. R. (2019). Adaptive F-actin polymerization and localized ATP production drive basement membrane invasion in the absence of MMPs. *Developmental Cell, 48*, 313–328.

Kessenbrock, K., Wang, C.-Y., & Werb, Z. (2015). Matrix metalloproteinases in stem cell regulation and cancer. *Matrix Biology, 44–46*, 184–190. https://doi.org/10.1016/j.matbio.2015.01.022.

Khalilgharibi, N., & Mao, Y. (2021). To form and function: On the role of basement membrane mechanics in tissue development, homeostasis and disease. *Open Biology, 11*. https://doi.org/10.1098/RSOB.200360.

Klotzsch, E., Smith, M. L., Kubow, K. E., Muntwyler, S., Little, W. C., Beyeler, F., ... Vogel, V. (2009). Fibronectin forms the most extensible biological fibers displaying switchable force-exposed cryptic binding sites. *Proceedings of the National Academy of Sciences of the United States of America, 106*, 18267–18272. https://doi.org/10.1073/pnas.0907518106.

Koser, D. E., Thompson, A. J., Foster, S. K., Dwivedy, A., Pillai, E. K., Sheridan, G. K., ... Franze, K. (2016). Mechanosensing is critical for axon growth in the developing brain. *Nature Neuroscience, 19*, 1592–1598. https://doi.org/10.1038/nn.4394.

Kyprianou, C., Christodoulou, N., Hamilton, R. S., Nahaboo, W., Boomgaard, D. S., Amadei, G., ... Zernicka-Goetz, M. (2020). Basement membrane remodelling regulates mouse embryogenesis. *Nature, 582*, 253–258. https://doi.org/10.1038/s41586-020-2264-2.

Lämmermann, T., Bader, B. L., Monkley, S. J., Worbs, T., Wedlich-Söldner, R., Hirsch, K., ... Fässler, R. (2008). Rapid leukocyte migration by integrin-independent flowing and squeezing. *Nature, 453*, 51–55.

Larreta-Garde, V., & Berry, H. (2002). Modeling extracellular matrix degradation balance with proteinase/transglutaminase cycle. *Journal of Theoretical Biology, 217*, 105–124. https://doi.org/10.1006/jtbi.2002.3010.

Le, H. A., & Mayor, R. (2023). Cell–matrix and cell–cell interaction mechanics in guiding migration. *Biochemical Society Transactions, 51*, 1733–1745. https://doi.org/10.1042/BST20230211.

Lee, H., Alisafaei, F., Adebawale, K., Chang, J., Shenoy, V. B., & Chaudhuri, O. (2021). The nuclear piston activates mechanosensitive ion channels to generate cell migration paths in confining microenvironments. *Science Advances, 7*, eabd4058. https://doi.org/10.1126/sciadv.abd4058.

Lee, H.-T., Huang, C.-H., Chen, W.-C., Tsai, C.-S., Chao, Y.-L., Liu, S.-H., ... Lee, Y.-J. (2018). Transglutaminase 2 promotes migration and invasion of lung cancer cells. *Oncology Research, 26*, 1175–1182. https://doi.org/10.3727/096504018X15149761920868.

Little, W. C., Smith, M. L., Ebneter, U., & Vogel, V. (2008). Assay to mechanically tune and optically probe fibrillar fibronectin conformations from fully relaxed to breakage. *Matrix Biology, 27*, 451–461. https://doi.org/10.1016/j.matbio.2008.02.003.

Liu, Y. J., Le Berre, M., Lautenschlaeger, F., Maiuri, P., Callan-Jones, A., Heuzé, M., ... Piel, M. (2015). Confinement and low adhesion induce fast amoeboid migration of slow mesenchymal cells. *Cell*. https://doi.org/10.1016/j.cell.2015.01.007.

Lo, C.-M., Wang, H.-B., Dembo, M., & Wang, Y. (2000). Cell movement is guided by the rigidity of the substrate. *Biophysical Journal, 79*, 144–152. https://doi.org/10.1016/S0006-3495(00)76279-5.

Lomakin, A., Cattin, C., Cuvelier, D., Alraies, Z., Molina, M., Nader, G., ... Zhitnyak, I. (2020). The nucleus acts as a ruler tailoring cell responses to spatial constraints. *Science (New York, N. Y.), 370*, eaba2894. https://doi.org/10.1126/science.aba2894.

Martinson, W. D., McLennan, R., Teddy, J. M., McKinney, M. C., Davidson, L. A., Baker, R. E., ... Maini, P. K. (2023). Dynamic fibronectin assembly and remodeling by leader neural crest cells prevents jamming in collective cell migration. *Elife, 12*, e83792.

Matsubayashi, Y. (2022). Dynamic movement and turnover of extracellular matrices during tissue development and maintenance. *Fly, 16*, 248–274. https://doi.org/10.1080/19336934.2022.2076539.

Matsubayashi, Y., Louani, A., Dragu, A., Sanchez-Sanchez, B. J., Serna-Morales, E., Yolland, L., ... Heddleston, J. M. (2017). A moving source of matrix components is essential for de novo basement membrane formation. *Current Biology, 27*, 3526–3534.

Matsubayashi, Y., Sánchez-Sánchez, B. J., Marcotti, S., Serna-Morales, E., Dragu, A., Díaz-de-la-Loza, M.-C., ... Stramer, B. M. (2020). Rapid homeostatic turnover of embryonic ECM during tissue morphogenesis. *Developmental Cell, 54*, 33–42.e9. https://doi.org/10.1016/j.devcel.2020.06.005.

McGregor, A. L., Hsia, C.-R., & Lammerding, J. (2016). Squish and squeeze—The nucleus as a physical barrier during migration in confined environments. *Current Opinion in Cell Biology, 40*, 32–40. https://doi.org/10.1016/j.ceb.2016.01.011.

Melamed, S., Zaffryar-Eilot, S., Nadjar-Boger, E., Aviram, R., Zhao, H., Yaseen-Badarne, W., ... Wolfenson, H. (2023). Initiation of fibronectin fibrillogenesis is an enzyme-dependent process. *Cell Reports, 42*, 112473. https://doi.org/10.1016/j.celrep.2023.112473.

Mitchison, T., & Kirschner, M. (1988). Cytoskeletal dynamics and nerve growth. *Neuron, 1*, 761–772.

Mohamed, F. F., Ge, C., Hallett, S. A., Bancroft, A. C., Cowling, R. T., Ono, N., ... Kaartinen, V. M. (2023). Control of craniofacial development by the collagen receptor, discoidin domain receptor 2. *Elife, 12*, e77257.

Molina López, E., Kabanova, A., Winkel, A., Franze, K., Palacios, I. M., & Martín-Bermudo, M. D. (2023). Constriction imposed by basement membrane regulates developmental cell migration. *PLoS Biology, 21*, e3002172. https://doi.org/10.1371/journal.pbio.3002172.

Morrissey, M. A., Keeley, D. P., Hagedorn, E. J., McClatchey, S. T. H., Chi, Q., Hall, D. H., & Sherwood, D. R. (2014). B-LINK: A hemicentin, plakin, and integrin-dependent adhesion system that links tissues by connecting adjacent basement membranes. *Developmental Cell, 31*, 319–331. https://doi.org/10.1016/j.devcel.2014.08.024.

Münster, S., Jawerth, L. M., Leslie, B. A., Weitz, J. I., Fabry, B., & Weitz, D. A. (2013). Strain history dependence of the nonlinear stress response of fibrin and collagen networks. *Proceedings of the National Academy of Sciences of the United States of America, 110*, 12197–12202. https://doi.org/10.1073/pnas.1222787110.

Naba, A. (2023). Ten years of extracellular matrix proteomics: Accomplishments, challenges, and future perspectives. *Molecular & Cellular Proteomics, 22*, 100528. https://doi.org/10.1016/j.mcpro.2023.100528.

Naba, A. (2024). Mechanisms of assembly and remodelling of the extracellular matrix. *Nature Reviews. Molecular Cell Biology, 25*, 865–885. https://doi.org/10.1038/s41580-024-00767-3.

Naba, A., Clauser, K. R., Hoersch, S., Liu, H., Carr, S. A., & Hynes, R. O. (2012). The matrisome: In silico definition and in vivo characterization by proteomics of normal and tumor extracellular matrices. *Molecular & Cellular Proteomics, 11*, M111.014647. https://doi.org/10.1074/mcp.M111.014647.

Nam, S., Hu, K. H., Butte, M. J., & Chaudhuri, O. (2016). Strain-enhanced stress relaxation impacts nonlinear elasticity in collagen gels. *Proceedings of the National Academy of Sciences of the United States of America, 113*, 5492–5497. https://doi.org/10.1073/pnas.1523906113.

Ohashi, T., Kiehart, D. P., & Erickson, H. P. (1999). Dynamics and elasticity of the fibronectin matrix in living cell culture visualized by fibronectin–green fluorescent protein. *Proceedings of the National Academy of Sciences of the United States of America, 96*, 2153–2158. https://doi.org/10.1073/pnas.96.5.2153.

Ortega, N., & Werb, Z. (2002). New functional roles for non-collagenous domains of basement membrane collagens. *Journal of Cell Science, 115*, 4201–4214. https://doi.org/10.1242/jcs.00106.

Pally, D., & Naba, A. (2024). Extracellular matrix dynamics: A key regulator of cell migration across length-scales and systems. *Current Opinion in Cell Biology, 86*, 102309. https://doi.org/10.1016/j.ceb.2023.102309.

Panayotou, G., End, P., Aumailley, M., Timpl, R., & Engel, J. (1989). Domains of laminin with growth-factor activity. *Cell, 56*, 93–101. https://doi.org/10.1016/0092-8674(89)90987-2.

Pankov, R., & Yamada, K. M. (2002). Fibronectin at a glance. *Journal of Cell Science, 115*, 3861–3863. https://doi.org/10.1242/jcs.00059.

Petrie, R. J., Harlin, H. M., Korsak, L. I. T., & Yamada, K. M. (2017). Activating the nuclear piston mechanism of 3D migration in tumor cells. *Journal of Cell Biology*. https://doi.org/10.1083/jcb.201605097.

Petrie, R. J., Koo, H., & Yamada, K. M. (2014). Generation of compartmentalized pressure by a nuclear piston governs cell motility in a 3D matrix. *Science (New York, N. Y.), 345*, 1062–1065. https://doi.org/10.1126/science.1256965.

Petrov, P. B., Considine, J. M., Izzi, V., & Naba, A. (2023). Matrisome AnalyzeR – A suite of tools to annotate and quantify ECM molecules in big datasets across organisms. *Journal of Cell Science, 136*, jcs261255. https://doi.org/10.1242/jcs.261255.

Piersma, B., Hayward, M.-K., & Weaver, V. M. (2020). Fibrosis and cancer: A strained relationship. *Biochimica et Biophysica Acta (BBA) – Reviews on Cancer, 1873*, 188356. https://doi.org/10.1016/j.bbcan.2020.188356.

Pozzi, A., Yurchenco, P. D., & Iozzo, R. V. (2017). The nature and biology of basement membranes. *Matrix Biology, 57–58*, 1–11. https://doi.org/10.1016/j.matbio.2016.12.009.

Prevedel, R., Diz-Muñoz, A., Ruocco, G., & Antonacci, G. (2019). Brillouin microscopy: An emerging tool for mechanobiology. *Nature Methods, 16*, 969–977. https://doi.org/10.1038/s41592-019-0543-3.

Rekad, Z., Izzi, V., Lamba, R., Ciais, D., & Van Obberghen-Schilling, E. (2022). The alternative matrisome: Alternative splicing of ECM proteins in development, homeostasis and tumor progression. *Matrix Biology, 111*, 26–52. https://doi.org/10.1016/j.matbio.2022.05.003.

Renkawitz, J., Kopf, A., Stopp, J., de Vries, I., Driscoll, M. K., Merrin, J., ... Fiolka, R. (2019). Nuclear positioning facilitates amoeboid migration along the path of least resistance. *Nature, 568*, 546–550. https://doi.org/10.1038/s41586-019-1087-5.

Ricard-Blum, S. (2011). The collagen family. *Cold Spring Harbor Perspectives in Biology, 3*, a004978. https://doi.org/10.1101/cshperspect.a004978.

Ruprecht, V., Wieser, S., Callan-Jones, A., Smutny, M., Morita, H., Sako, K., ... Heisenberg, C. P. (2015). Cortical contractility triggers a stochastic switch to fast amoeboid cell motility. *Cell*. https://doi.org/10.1016/j.cell.2015.01.008.

Sadati, M., Taheri Qazvini, N., Krishnan, R., Park, C. Y., & Fredberg, J. J. (2013). Collective migration and cell jamming. *Differentiation; Research in Biological Diversity, 86*, 121–125. https://doi.org/10.1016/j.diff.2013.02.005.

Saraswathibhatla, A., Indana, D., & Chaudhuri, O. (2023). Cell–extracellular matrix mechanotransduction in 3D. *Nature Reviews. Molecular Cell Biology, 24*, 495–516.

Scarpa, E., & Mayor, R. (2016). Collective cell migration in development. *Journal of Cell Biology*. https://doi.org/10.1083/jcb.201508047.

Scarpellini, A., Germack, R., Lortat-Jacob, H., Muramatsu, T., Billett, E., Johnson, T., & Verderio, E. A. M. (2009). Heparan sulfate proteoglycans are receptors for the cell-surface trafficking and biological activity of transglutaminase-2. *Journal of Biological Chemistry, 284*, 18411–18423. https://doi.org/10.1074/jbc.M109.012948.

Scott, J. E. (2003). Elasticity in extracellular matrix 'shape modules' of tendon, cartilage, etc. A sliding proteoglycan-filament model. *The Journal of Physiology, 553*, 335–343.

Sekiguchi, R., & Yamada, K. M. (2018). *Basement membranes in development and disease. Current topics in developmental biology*. Elsevier, 143–191. https://doi.org/10.1016/bs.ctdb.2018.02.005.

Senk, A., & Djonov, V. (2021). Collagen fibers provide guidance cues for capillary regrowth during regenerative angiogenesis in zebrafish. *Scientific Reports, 11*, 19520.

Serwane, F., Mongera, A., Rowghanian, P., Kealhofer, D. A., Lucio, A. A., Hockenbery, Z. M., & Campàs, O. (2017). In vivo quantification of spatially varying mechanical properties in developing tissues. *Nature Methods, 14*, 181–186. https://doi.org/10.1038/nmeth.4101.

Shao, X., Gomez, C. D., Kapoor, N., Considine, J. M., Grams, C., Gao, Y. (Tom), & Naba, A. (2023). MatrisomeDB 2.0: 2023 updates to the ECM-protein knowledge database. *Nucleic Acids Research, 51*, D1519–D1530. https://doi.org/10.1093/nar/gkac1009.

Shellard, A., & Mayor, R. (2021). Durotaxis: The hard path from in vitro to in vivo. *Developmental Cell, 56*, 227–239. https://doi.org/10.1016/j.devcel.2020.11.019.

Sherwood, D. R. (2021). Basement membrane remodeling guides cell migration and cell morphogenesis during development. *Current Opinion in Cell Biology, 72*, 19–27. https://doi.org/10.1016/j.ceb.2021.04.003.

Shimoda, M., Ohtsuka, T., Okada, Y., & Kanai, Y. (2021). Stromal metalloproteinases: Crucial contributors to the tumor microenvironment. *Pathology International, 71*, 1–14. https://doi.org/10.1111/pin.13033.

Siekhaus, D., Haesemeyer, M., Moffitt, O., & Lehmann, R. (2010). RhoL controls invasion and Rap1 localization during immune cell transmigration in Drosophila. *Nature Cell Biology, 12*, 605–610.

Soans, K. G., Ramos, A. P., Sidhaye, J., Krishna, A., Solomatina, A., Hoffmann, K. B., ... Modes, C. D. (2022). Collective cell migration during optic cup formation features changing cell-matrix interactions linked to matrix topology. *Current Biology, 32*, 4817–4831.

Spurlin, T. A., Bhadriraju, K., Chung, K.-H., Tona, A., & Plant, A. L. (2009). The treatment of collagen fibrils by tissue transglutaminase to promote vascular smooth muscle cell contractile signaling. *Biomaterials, 30*, 5486–5496. https://doi.org/10.1016/j.biomaterials.2009.07.014.

Sunyer, R., & Trepat, X. (2020). Durotaxis. *Current Biology, 30*, R383–R387. https://doi.org/10.1016/j.cub.2020.03.051.

Thiam, H.-R., Vargas, P., Carpi, N., Crespo, C. L., Raab, M., Terriac, E., ... Stradal, T. (2016). Perinuclear Arp2/3-driven actin polymerization enables nuclear deformation to facilitate cell migration through complex environments. *Nature Communications*, 7, 10997.

Thomas, D. G., Yenepalli, A., Denais, C. M., Rape, A., Beach, J. R., Wang, Y. li, ... Egelhoff, T. T. (2015). Non-muscle myosin IIB is critical for nuclear translocation during 3D invasion. *Journal of Cell Biology*. https://doi.org/10.1083/jcb.201502039.

Thompson, A. J., Pillai, E. K., Dimov, I. B., Foster, S. K., Holt, C. E., & Franze, K. (2019). Rapid changes in tissue mechanics regulate cell behaviour in the developing embryonic brain. *eLife, 8*. https://doi.org/10.7554/elife.39356.

Töpfer, U. (2023). Basement membrane dynamics and mechanics in tissue morphogenesis. *Biology Open, 12*, bio059980. https://doi.org/10.1242/bio.059980.

Tran, K. T., Lamb, P., & Deng, J.-S. (2005). Matrikines and matricryptins: Implications for cutaneous cancers and skin repair. *Journal of Dermatological Science, 40*, 11–20. https://doi.org/10.1016/j.jdermsci.2005.05.001.

Trier, J. S., Allan, C. H., Abrahamson, D. R., & Hagen, S. J. (1990). Epithelial basement membrane of mouse jejunum. Evidence for laminin turnover along the entire crypt-villus axis. *The Journal of Clinical Investigation, 86*, 87–95. https://doi.org/10.1172/JCI114720.

Van De Bor, V., Loreau, V., Malbouyres, M., Cerezo, D., Placenti, A., Ruggiero, F., & Noselli, S. (2021). A dynamic and mosaic basement membrane controls cell intercalation in *Drosophila* ovaries. *Development (Cambridge, England), 148*, dev195511. https://doi.org/10.1242/dev.195511.

Ventura, G., & Sedzinski, J. (2022). Emerging concepts on the mechanical interplay between migrating cells and microenvironment in vivo. *Frontiers in Cell and Developmental Biology, 10*, 961460. https://doi.org/10.3389/fcell.2022.961460.

Venturini, V., Pezzano, F., Catala Castro, F., Häkkinen, H.-M., Jiménez-Delgado, S., Colomer-Rosell, M., ... Valverde, M. A. (2020). The nucleus measures shape changes for cellular proprioception to control dynamic cell behavior. *Science (New York, N. Y.), 370*, eaba2644.

Walker, C., Mojares, E., & Del Río Hernández, A. (2018). Role of extracellular matrix in development and cancer progression. *IJMS, 19*, 3028. https://doi.org/10.3390/ijms19103028.

Walma, D. A. C., & Yamada, K. M. (2020). The extracellular matrix in development. *Development, 147*, dev175596.

Wang, W. Y., Davidson, C. D., Lin, D., & Baker, B. M. (2019). Actomyosin contractility-dependent matrix stretch and recoil induces rapid cell migration. *Nature Communications, 10*, 1186.

Wang, S., Matsumoto, K., Lish, S. R., Cartagena-Rivera, A. X., & Yamada, K. M. (2021). Budding epithelial morphogenesis driven by cell-matrix versus cell-cell adhesion. *Cell, 184*, 3702–3716.

Wang, W. Y., Pearson, A. T., Kutys, M. L., Choi, C. K., Wozniak, M. A., Baker, B. M., & Chen, C. S. (2018). Extracellular matrix alignment dictates the organization of focal adhesions and directs uniaxial cell migration. *APL Bioengineering, 2*.

Wolf, K., Mazo, I., Leung, H., Engelke, K., Von Andrian, U. H., Deryugina, E. I., ... Friedl, P. (2003). Compensation mechanism in tumor cell migration: Mesenchymal-amoeboid transition after blocking of pericellular proteolysis. *Journal of Cell Biology*. https://doi.org/10.1083/jcb.200209006.

Wolf, K., te Lindert, M., Krause, M., Alexander, S., te Riet, J., Willis, A. L., ... Friedl, P. (2013). Physical limits of cell migration: Control by ECM space and nuclear deformation and tuning by proteolysis and traction force. *Journal of Cell Biology*. https://doi.org/10.1083/jcb.201210152.

Wu, D., Yamada, K. M., & Wang, S. (2023). Tissue morphogenesis through dynamic cell and matrix interactions. *Annual Review of Cell and Developmental Biology, 39*, 123–144. https://doi.org/10.1146/annurev-cellbio-020223-031019.

Yamada, K. M., Collins, J. W., Cruz Walma, D. A., Doyle, A. D., Morales, S. G., Lu, J., ... Wang, S. (2019). Extracellular matrix dynamics in cell migration, invasion and tissue morphogenesis. *International Journal of Experimental Pathology, 100*, 144–152. https://doi.org/10.1111/iep.12329.

Yamada, K. M., Doyle, A. D., & Lu, J. (2022). Cell–3D matrix interactions: Recent advances and opportunities. *Trends in Cell Biology, 32*, 883–895. https://doi.org/10.1016/j.tcb.2022.03.002.

Yamada, K. M., & Sixt, M. (2019). Mechanisms of 3D cell migration. *Nature Reviews. Molecular Cell Biology, 20*, 738–752. https://doi.org/10.1038/s41580-019-0172-9.

Yamaguchi, N., Zhang, Z., Schneider, T., Wang, B., Panozzo, D., & Knaut, H. (2022). Rear traction forces drive adherent tissue migration in vivo. *Nature Cell Biology, 24*, 194–204.

Yolland, L., Burki, M., Marcotti, S., Luchici, A., Kenny, F. N., Davis, J. R., ... Davidson, A. (2019). Persistent and polarized global actin flow is essential for directionality during cell migration. *Nature Cell Biology, 21*, 1370–1381.

Yurchenco, P. D. (2011). Basement membranes: Cell scaffoldings and signaling platforms. *Cold Spring Harbor Perspectives in Biology, 3*, a004911. https://doi.org/10.1101/cshperspect.a004911.

Zajac, A. L., & Horne-Badovinac, S. (2022). Kinesin-directed secretion of basement membrane proteins to a subdomain of the basolateral surface in Drosophila epithelial cells. *Current Biology, 32*, 735–748.e10. https://doi.org/10.1016/j.cub.2021.12.025.

Zamir, E. A., Czirók, A., Cui, C., Little, C. D., & Rongish, B. J. (2006). Mesodermal cell displacements during avian gastrulation are due to both individual cell-autonomous and convective tissue movements. *Proceedings of the National Academy of Sciences of the United States of America, 103*, 19806–19811. https://doi.org/10.1073/pnas.0606100103.

Zamir, E. A., Rongish, B. J., & Little, C. D. (2008). The ECM moves during primitive streak formation—Computation of ECM versus cellular motion. *PLoS Biology, 6*, e247. https://doi.org/10.1371/journal.pbio.0060247.

Zhang, T., Yin, C., Qiao, L., Jing, L., Li, H., Xiao, C., ... Zhu, H. (2014). Stat3-Efemp2a modulates the fibrillar matrix for cohesive movement of prechordal plate progenitors. *Development (Cambridge, England), 141*, 4332–4342.

Zhu, M., Tao, H., Samani, M., Luo, M., Wang, X., Hopyan, S., & Sun, Y. (2020). Spatial mapping of tissue properties in vivo reveals a 3D stiffness gradient in the mouse limb bud. *Proceedings of the National Academy of Sciences of the United States of America, 117*, 4781–4791. https://doi.org/10.1073/pnas.1912656117.

CHAPTER THREE

From neural crest migration to the onset of gangliogenesis

Hugo A. Urrutia and Marianne E. Bronner*
Division of Biology and Biological Engineering, California Institute of Technology, Pasadena, CA, United States
*Corresponding author. e-mail address: mbronner@caltech.edu

Contents

1. Introduction 68
2. Neural crest induction and formation 72
3. Neural crest EMT and onset of migration 74
4. Cranial neural crest migration 75
   4.1 Trunk neural crest migration 80
5. Onset of cranial neural crest ectomesenchymal differentiation 82
   5.1 Cranial neural crest and placode contribution to peripheral nervous systems 84
   5.2 Epibranchial ganglia 87
   5.3 Trigeminal ganglia 88
6. Concluding remarks 91
Acknowledgments 92
References 92

Abstract

The neural crest is a highly migratory and multipotent cell population that contributes to many defining features of vertebrates. As a uniquely vertebrate cell type, the neural crest is an excellent model for studying cell lineage and diversification during embryonic development because of its multipotency, motility, and capacity to form a plethora of derivatives. Neural crest cells migrate extensively throughout the body and contribute to many of the defining features of vertebrate embryos, including the craniofacial skeleton, most of the peripheral nervous system and pigmentation of the skin. What guides their migration and subsequent formation of discrete structures? Interactions between neural crest cells and their environment, including other cell types like placode cells, play a major role in guiding their migration and condensation into numerous derivatives. In this review, we discuss aspects of neural crest induction, migration and axial level differences, highlighting what is currently known regarding molecular cues that govern their formation, migratory behavior, and differentiation as they reach their final destinations. We particularly focus on formation of cranial sensory ganglia. New technologies are playing an important role in furthering our understanding of the molecular mechanisms underlying neural crest migration and what leads to cessation of their movement and onset of differentiation.

Current Topics in Developmental Biology, Volume 164
ISSN 0070-2153, https://doi.org/10.1016/bs.ctdb.2025.02.003

1. Introduction

The neural crest (NC) is a transient, highly migratory and multipotent embryonic cell population, specific to vertebrates. Often referred to as the fourth germ layer, the neural crest originates from the ectoderm but has broad developmental potential to differentiate into a wide variety of distinct cell types, including ectomesenchymal cell types of the face and most of the peripheral nervous system (Hall, 1998, 2000, 2008; N. Le Douarin & Kalcheim, 1999; N.M. Le Douarin & Dupin, 2014; Shyamala, Yanduri, Girish, & Murgod, 2015). The unique capabilities of the neural crest have led to the "new head hypothesis", which proposes that acquisition of the neural crest together with ectodermal placodes in ancestral vertebrates enabled expansion of the brain which in turn aided predation (Gans & Northcutt, 1983; Glenn Northcutt, 2005).

Induction of the NC initiates during gastrulation at the border of the neural plate, juxtaposed between neural and non-neural ectoderm (Fig. 1A). As the embryo develops, the neural crest cells undergo specification within the neural plate border, initially residing within the elevating neural folds during neurulation and subsequently positioned within the dorsal neural tube (Fig. 1A, B and C) (Erickson & Reedy, 1998; Milet & Monsoro-Burq, 2012; Steventon et al., 2005). By the time these cells are fully specified to a neural crest fate, they express characteristic neural crest marker genes – *FoxD3*, *Snai2*, and *Sox9*, among others. Upon closure of the neural tube, avian NC cells undergo an epithelial-to-mesenchymal transition (EMT), whereby they lose their epithelial characteristics, such as apicobasal polarity and tight junctions, and assume invasive and migratory mesenchymal characteristics, thus becoming a mesenchymal population distinct from the neural tube or non-neural ectoderm (Fig. 1C) (Duband & Thiery, 1982; Hall, 2009; Kerosuo & Bronner-Fraser, 2012; Le Douarin & Kalcheim, 1999; Lim & Thiery, 2012; Sauka-Spengler & Bronner-Fraser, 2008; Simoes-Costa & Bronner, 2016; Steventon et al., 2005; Theveneau & Mayor, 2012; Théveneau et al., 2007; Thiery & Sleeman, 2006).

Following EMT and delamination from the neuroepithelium, NC cells emigrate from the dorsal neural tube, either collectively or individually. The process of neural tube closure initiates in the head and proceeds tailward, as the embryo develops. Following this rostral to caudal sequence, neural crest emigration extends along the body axis in a rostrocaudal wave (Fig. 1D). Subsequently, NC cells migrate extensively throughout the embryo along well-defined routes (Fig. 1D, E and F), responding to local

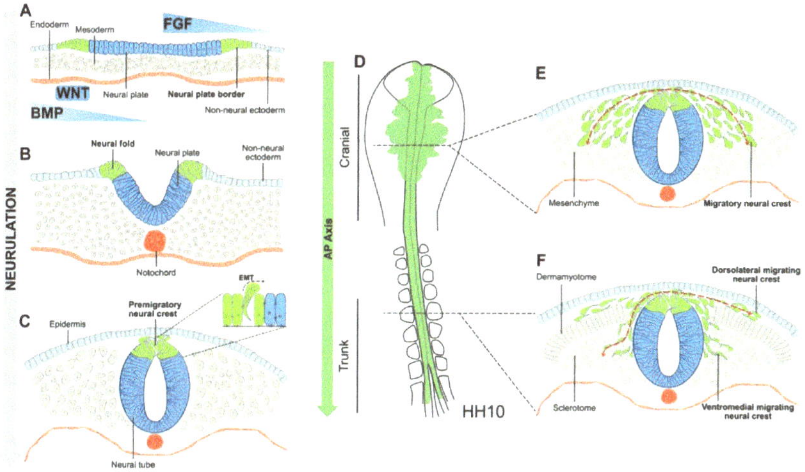

Fig. 1 Generalized morphogenetic stages during early neural crest development. (A) Schematic diagram of transverse sections through chick embryo during neurulation. Prospective neural crest (NC) cells reside in the neural plate border (green), a territory between the neural plate and the non-neural ectoderm. NC is induced by signals (mainly BMPs, WNT, and/or FGs) at the neural plate border. (B) As neurulation proceeds, the neural plate invaginates, resulting in the elevation of the neural folds, which contain NC precursors. (C) After neural tube closure, NC cells lose intercellular connections and undergo an epithelial to mesenchymal transition (EMT). (D) Schematic dorsal view of a stage 10 chicken embryo, showing the NC (green) in the vicinity of the midline. Dotted lines delimit the embryonic region represented in cross-section (E and F). (E and F) Schematic of migratory NC cells following stereotypical pathways to diverse destinations. HH: Hamburger and Hamilton developmental stages, AP: anteroposterior.

environmental and morphogenetic cues, enabling their invasion of distinct tissues, where they will differentiate into a wide range of cell types (Fig. 2) (Dupin et al., 2006; Grenier et al., 2009; Le Douarin & Teillet, 1974; Minoux & Rijli, 2010; Theveneau & Mayor, 2012).

Pioneering work by Le Douarin and colleagues has shown that avian neural crest populations differ in developmental potential along the rostrocaudal (anteroposterior) body axis, such that they can be sub-categorized into four distinct regions, based on their axial level of origin (Fig. 2). For example, the cranial NC, which arises from the cranial neural tube (posterior forebrain to the anterior hindbrain) migrate dorsolaterally between the presumptive epidermis and the mesodermal mesenchyme (Fig. 1E) to populate the head, forming the craniofacial skeleton and contributing to cranial sensory ganglia (Fig. 2) (Johnston, 1966; Noden, 1975). The vagal neural crest contributes to portions of the heart as well

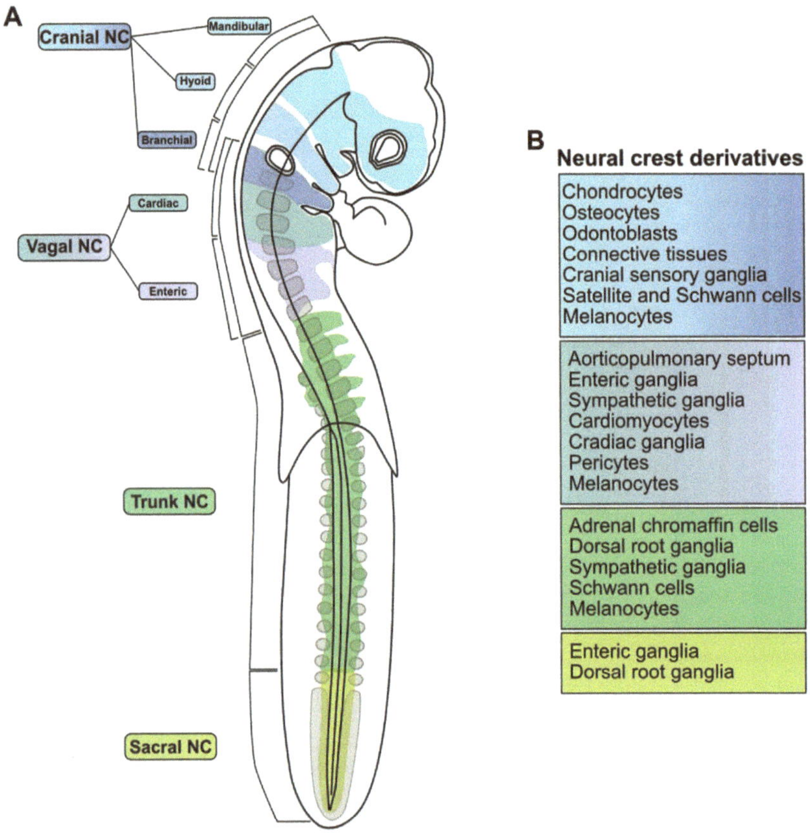

Fig. 2 Overview of NC subpopulations along the anteroposterior axis. (A) Schematic of NC subpopulations in an HH16 chick embryo.' populations are subdivided into several unique subpopulations along the body axis, designated on their axial level of origin - cranial, vagal, trunk, and sacral NC. The cranial NC includes NC cells that are formed between the forebrain and the 6th rhombomere of the hindbrain. The vagal crest is emerge from the first seven somites. The trunk neural crest spans from the 8th to the 27th somite, while the sacral neural crest is positioned posterior to this boundary. (B) Major derivatives of NC subpopulations.

as the enteric ganglia and, in birds, the nerve of Remak of the gut (Johnston, 1966; Noden, 1975). The vagal NC population arises from the posterior hindbrain to somite 7 (Fig. 2). Just below the vagal region is the trunk NC, arising from the presumptive spinal cord adjacent to somites 8–28. These NC cells contribute to the dorsal root ganglion, sympathetic ganglia, and the adrenal medulla (Fig. 2) (Dupin et al., 2006; Grenier et al., 2009; Le Douarin & Teillet, 1974; Minoux & Rijli, 2010;

Theveneau & Mayor, 2012). Finally, the lumbosacral neural crest arise posterior to the 28th somite and contributes to the enteric nervous system of the post-umbilical gut.

The best studied of these neural crest subpopulations is the cranial (cephalic) neural crest (CNC) which arises from the dorsal side of the diencephalon (posterior forebrain) extending to the caudal rhombencephalon (hindbrain) in avian embryos (Figs. 2 and 3) (Couly & Le Douarin, 1987). After delaminating, these cells predominantly migrate as one continuous wave from the dorsal midline of the neural tube, and then split into three discrete, segregated streams: first, ranging from the diencephalon to rhombomeres (r) 1 and r2; second, laterally from r4; third, laterally from r6-r8 (Fig. 3). These streams populate specific regions of the developing craniofacial and neck region. In particular, the first cranial-most (mandibular) stream populates the frontal-nasal and periocular regions, along with the first branchial (pharyngeal) arch (BA) (Fig. 3). The second (hyoid) stream emerging from r4 contributes extensively to BA2, while the third (branchial) stream originating from r6 populates BA3, the heart and enteric nervous system (Fig. 3) (Le Douarin & Kalcheim, 1999).

In this review, we discuss recent insights into the molecular cues that govern and direct neural crest formation, migratory behaviors, and initial differentiation as they reach their final destinations. While we primarily focus on the cranial neural crest, because it is best studied, we also discuss

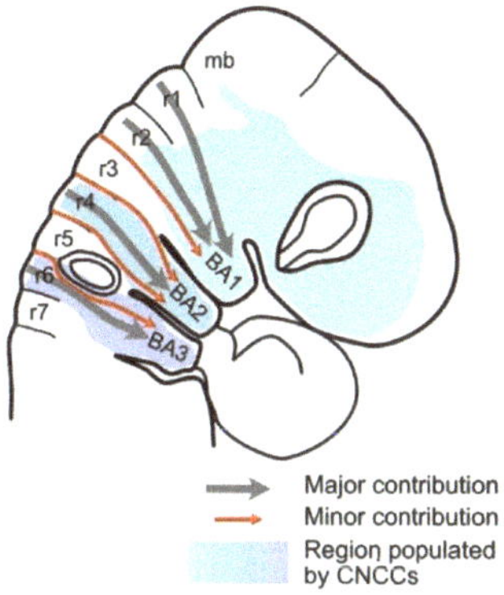

Fig. 3 Overview of cranial neural crest migratory route. Beginning at stage 10, cranial NC cells migrate dorsolaterally, between the ectoderm and the underlying paraxial mesoderm. The major migratory streams of the cranial NC originate from rhombomeres 2, 4 and 6 (r2, r4), (BAI, BA2, BA), resputively. Some C all rom. 23 and35 (red arrows) contribute minimally to BA1, BA2 and BA3. mb: midbrain.

differences between cranial and trunk neural crest migration. Finally, we present an overview of their contributions and the final derivates to which these cells give rise to, largely focusing on their contribution to cranial sensory gangliogenesis.

2. Neural crest induction and formation

The neural crest was first described in the literature over 150 years ago, when Wilhelm His, Sr. identified this strand of cells emerging from the dorsal neural tube. Initially, His coined the term *Zwischenstrang* (the intermediate cord) that was later referred to as the neural crest (Baker, 2008; Bronner & LeDouarin, 2012; Couly et al., 1993; His, 1868; Le Douarin & Dupin, 2018; Sauka-Spengler & Bronner-Fraser, 2008; Trainor et al., 2003; York & McCauley, 2020; Young et al., 2014). The ectoderm of gastrulating embryos will form four distinct cell types – epidermal cells, placodal cells, cells of the nervous system, and neural crest cells (Bronner-Fraser, 1995). Induction of the neural crest initiates at the neural plate border under the influence of extracellular signals like bone morphogenetic protein (BMP), fibroblast growth factor (FGF), Notch, retinoic acid, and WNT, which emanate from both the surrounding ectoderm and underlying mesoderm (Fig. 1A) (Betancur et al., 2010; Bronner & LeDouarin, 2012; Sauka-Spengler & Bronner-Fraser, 2008; Simões-Costa & Bronner, 2015; Thawani & Groves, 2020). Importantly, induction of the NC is governed by intermediate levels of BMP, controlled by BMP antagonists - Chordin, Noggin or Follistatin, originating from the mesoderm (Sauka-Spengler & Bronner-Fraser, 2008; Teraoka et al., 2009).

Neural tube closure (Fig. 1A, B, and C) progresses in an anteroposterior (rostrocaudal) sequence within the developing avian embryo (Fig. 1D), a progression closely followed by neural crest specification. In response to the signaling cascades, neural plate border and neural crest specifiers are activated. This includes various transcription factors such as *FoxD3*, *Pax3/7*, and *Tfap2*, which in turn further solidify NC formation (Betancur et al., 2010; Bronner & LeDouarin, 2012; Sauka-Spengler & Bronner-Fraser, 2008; Simões-Costa & Bronner, 2015). For instance, *FoxD3* has been implicated in the induction and specification of pre-migratory NC cells (Lukoseviciute et al., 2018). In addition, *Pax3/7* has been shown to be required for NC specification (Basch et al., 2006; Hong & Saint-Jeannet, 2007; Mayran et al., 2018). Conversely, the pioneer factor, *Tfap2* has been shown to affect chromatin landscape, further solidifying NC specification (Kenny et al., 2021; Wang et al., 2011). Nevertheless, combinatorial expression of various other transcription factors (e.g., *FoxD3*, *Id2*, *Pax3/7*, *Msx1/2*, *Sox9*, *Zic1/2*) within these territories also regulate NC specification and actively repressed alternative fates. For example, *FoxD3* mediates repression of neural and pre-placodal fates (Hochgreb-Hägele & Bronner, 2013; Lukoseviciute et al., 2018; Thomas & Erickson, 2009). Following specification, the NC are then positioned in the dorsal neural tube (Fig. 1C).

After induction and specification, the axial level-specific identity of NC cells is regulated by expression of different transcription factors. For instance, *Dmbx1* and *Otx2* play vital roles in specification of cranial NC identity, whereas anterior Hox genes, such as *HoxA2* and *HoxB2*, influence NC cells within the anterior hindbrain, while subsequent Hox genes (e.g. *HoxA3/4, HoxB3/4, HoxB5*), along with Retinoic Acid receptors and *Cdx1/2*, establish vagal and trunk NC identity (Frith et al., 2018; Minoux & Rijli, 2010; Rothstein et al., 2018; Simoes-Costa & Bronner, 2016; Williams et al., 2019). These signals activate expression of gene regulatory networks that define NC territories, and the subsequent steps required for NC development (Marchant et al., 1998; Mayor et al., 1995, 1997; Prasad et al., 2012; Sauka-Spengler & Bronner-Fraser, 2008; Steventon et al., 2009; Stuhlmiller & García-Castro, 2012). For example, ectopic expression of cranial crest-specific transcription factors (*Sox8*, *Tfap2β* and *Ets1*) in the trunk NC activates a partial reprogramming of avian trunk NC cells, resulting in formation of chondroblasts, a cell type that trunk neural crest cells normally are unable to make (Simoes-Costa & Bronner, 2016).

The neural plate border not only gives rise to NC cells, but also placodal precursors which arise adjacent and lateral to the neural crest in the pre-placodal domain (Betancur et al., 2010; Bhattacharyya et al., 2004; LaBonne & Bronner-Fraser, 1998; Pieper et al., 2011; Prasad et al., 2012; Streit, 2002; Xu et al., 2008). Whether presumptive neural crest and placode cells are predetermined within the neural plate border remains a matter of debate (Krispin et al., 2010; Roellig et al., 2017; Williams et al., 2022). The distinction between these cell types becomes apparent only at the end of neurulation when the midline of the ectoderm thickens and begins to invaginate, giving rise to the central nervous system (brain and spinal cord) along the body axis of the embryo (Fig. 1B) (Bronner-Fraser, 1995).

3. Neural crest EMT and onset of migration

Upon neural tube closure, pre-migratory neural crest cells undergo EMT, during which their transcriptional landscape and cellular properties change dramatically. They lose their epithelial characteristics and acquire mesenchymal properties to delaminate from the neural tube and migrate extensively throughout the developing embryo (Fig. 1D, E and F) (Betancur et al., 2010; Martik & Bronner, 2017; Martinsen & Bronner-Fraser, 1998; Minoux & Rijli, 2010; Minoux et al., 2017; Prasad et al., 2012; Sauka-Spengler & Bronner-Fraser, 2008; Simões-Costa & Bronner, 2015; Steventon et al., 2014; Theveneau & Mayor, 2012). Neural crest cells then emigrate from the dorsal neural tube in a rostrocaudal progression along the body axis with cells at the midline of the caudal forebrain and midbrain emerging prior to those adjacent to the hindbrain (Fig. 1D). Extracellular matrix (ECM) remodeling and loss of adherens junctions occur during EMT, via changes in cadherin expression. In particular, neural crest specification is complete when these cells switch from type I cadherins (E- and N-cadherin) and begin expressing type II cadherins (e.g. Cadherin-6B/7/10/11), which are essential for acquisition of contact inhibition of locomotion (CIL) and cell repolarization (Dady et al., 2012; Piloto & Schilling, 2010; Rogers et al., 2013; Scarpa et al., 2015; Théveneau et al., 2007; Theveneau et al., 2010).

Cranial NC cells have been shown to collectively migrate, which is defined either as coordinated migration of tightly clustered cells or small loose groups of cells, where communication and cooperation between these cells contribute to their overall directionality (Etienne-Manneville, 2014; Revenu et al., 2014; Theveneau & Mayor, 2012). Cadherin based contacts have been shown to control collective migration of cranial NC cells. These contacts act as key regulators of molecular communication, allowing cells to polarize, collectively migrate with high directionality, and coalesce upon reaching their destination (Carmona-Fontaine et al., 2008; Desai et al., 2009; Dupin et al., 2009; Friedl & Gilmour, 2009; Revenu et al., 2014; Theveneau et al., 2010).

EMT is controlled by NC specifier genes; e.g., in the trunk region, *Snai1/2*, *FoxD3*, and *Sox9/10* govern EMT. In contrast, these transcription factors work in conjunction with *Ets1*, *Sox5* and *p53* to mediate EMT in the cranial region (Cheung et al., 2005; del Barrio & Nieto, 2002; McKeown et al., 2013; Perez-Alcala et al., 2004; Rinon et al., 2011; Taneyhill et al., 2007; Théveneau et al., 2007). WNT signaling activates expression of *Snai1/2*, which is crucial for the initial delamination of NC cells via repression of cadherin molecules (Vallin et al., 2001; Yook et al., 2006). In *Xenopus*, cranial NC cells mainly express both type I E-cadherin and N-cadherin. At the onset of migration, E-cadherin expression is reduced while N-cadherin expression is only slightly reduced, through expression of the transcription factor, *Twist*. Similarly, expression of *Sip1/Zeb2* in chick is required for two phases of neural crest EMT. This transcription factor actively represses expression of E-cadherin in the presumptive NC, thereby allowing detachment from the neural tube, and activation and maintenance of N-cadherin expression during migration (Barriga et al., 2013; Rogers et al., 2013; Theveneau et al., 2013).

In chick, *Snai2* is essential for cranial NC EMT as it transcriptionally represses Cadherin-6B, mainly by functioning in conjunction with *Ets1* (Taneyhill et al., 2007; Théveneau et al., 2007). NC specifiers *FoxD3* and *Sox10* repress N-cadherin expression (Cheung et al., 2005; Dottori et al., 2001). In addition, *FoxD3* has been shown to further contribute to EMT by repressing other transmembrane proteins, like *Tspan18*, which is crucial for establishment and maintenance of tight cell junctions (Fairchild & Gammill, 2013; Fairchild et al., 2014). Conversely, as expression of type I cadherins is reduced, type II cadherins (e.g. Cadherin-7/11/19/20) are upregulated in migratory NC (Chalpe et al., 2010; Faulkner-Jones et al., 1999; Moore & Larue, 2004; Nakagawa & Takeichi, 1995; Pla et al., 2001; Takahashi & Osumi, 2005; Wu & Taneyhill, 2019; Wu et al., 2014). Finally, upon completion of EMT, NC cells begin receiving and integrating both positive and negative signals from surrounding tissues, via cell-cell and cell-ECM contact interactions, allowing these cells to migrate extensively, exhibiting both collective and individual migratory behaviors (Hall, 2008; Le Douarin & Kalcheim, 1999).

4. Cranial neural crest migration

Although the general sequence of induction, delamination, and initiation of migration appear similar along most of the anteroposterior axis, NC migration and the range of derivatives vary among different axial regions and in different species (Alfandari et al., 2010; Hall, 2009; Le Douarin & Kalcheim, 1999; Minoux & Rijli, 2010; Théveneau et al., 2007). Various lineage tracing methods marking different populations of NC cells have significantly enhanced our understanding of the migratory paths and derivatives these cells form.

In avian embryos, advent of the quail-chick chimeras by Le Douarin and colleagues (Le Douarin & Teillet, 1973, 1974; Le Douarin, 1973) enabled precise mapping of neural crest migratory paths, axial levels of origin, and range of derivatives. Neural crest migration has also been extensively studied in *Xenopus* and zebrafish embryos. In frogs, cranial neural crest cells undergo collective migration, whereas in chick and zebrafish, they migrate more as individual cells that have leader and follower cells (Kulesa & Gammill, 2010; Kuriyama & Mayor, 2008; Olsson & Hanken, 1996; Richardson et al., 2016; Schroeder, 1970). Tracing neural crest cell behavior by both static and live imaging has provided invaluable insight into their migratory patterns (Davis & Trinkaus, 1981; Le Douarin & Kalcheim, 1999; Newgreen et al., 1982). Moreover, comparative studies across different species have highlighted the complexity of cranial NC migratory behaviors, the nature of their migratory streams with respect to individual or collective cell migration, and the role of environmental cues in providing attractive and repulsive signals (Bronner-Fraser, 1993; Erickson et al., 1980; Erickson, 1985; Le Douarin, 1982; Newgreen, 1989; Noden, 1975; Thiery et al., 1982; Tosney, 1982).

Cranial neural crest cells emerge from the length of the posterior forebrain (diencephalon) to the rostral hindbrain (rhombencephalon) (Figs. 2 and 3). These cells predominantly migrate along a dorsolateral route, under the presumptive ectoderm (Fig. 1E), and eventually occupy the craniofacial/frontonasal region and branchial arches (BA) of the developing avian embryo (Fig. 3) (Bajpai et al., 2010; Creuzet et al., 2002; Johnston, 1966; Kerosuo & Bronner, 2016; Kulesa et al., 2010; Le Douarin, 1973; Noden, 1975; Weston & Butler, 1966; Weston, 1963). Cranial NC migrate in three distinct streams. The mandibular stream emerges from the level of the posterior forebrain to rhombomere (r) 2, with a small contribution from the adjacent region, r3, which is predominantly NC-free (Fig. 3). The cells in this stream migrate dorsolaterally for the initial portion of their migration to populate the frontal-nasal and periocular regions, along with BA1 (Fig. 3). The hyoid stream of cranial NC cells arises from r4 and eventually populates BA2, with neural crest cells migrating downward from r3 and upward from r5 contributing small portions to this stream (Fig. 3). The branchial cranial NC stream originates from r6–r8, with contribution from both r5 and r7, and populates BA3 and BA4 (Fig. 3) (Kuriyama & Mayor, 2008; Le Douarin & Kalcheim, 1999). In part, the main reason for these well-segregated streams is the lack of NC cells originating and migrating from r3 and r5, due to the existence of focal apoptosis mainly mediated by BMP4, although some aspects of this proposed signaling cascade have been challenged (Farlie et al., 1999; Lumsden et al., 1994; Smith & Graham, 2001). BMP receptors BMPR1A and BMPR2 are expressed in both the hindbrain and NC cells that arise from these axial levels, but BMP4 is only expressed at the level of r3 and r5. Studies have shown that when r4 NC cells are transplanted to either r3 or r5, they failed to properly migrate, suggesting that these spaces are devoid of NC due to environmental cues rather than cell-autonomous effects (Farlie et al., 1999). *In vivo* time-lapses have shown that NC cells emerging from r3 or r5 rarely migrate, instead rounding up and collapsing their filopodia and lamellipodia, and reversing their migration to join adjacent NC streams (Kulesa & Fraser, 1998, 2000).

The combination of high resolution live imaging using vital dyes or transgenics together with molecular/genetic perturbations, has illuminated the molecular mechanisms governing the migratory behavior of NC cells. Largely focusing on the cranial neural crest in chick, *Xenopus*, and zebrafish embryos, several studies have shown that the cranial NC navigates through complex environments. Migrating neural crest cells are influenced by factors secreted by neighboring cells including the cranial placodes, rhombomeres and surrounding mesenchyme (Baker, 2008; Minoux & Rijli, 2010; Steventon et al., 2014; Szabó et al., 2019; Theveneau & Mayor, 2012).

Three main mechanisms provide guidance to the cranial neural crest as they navigate through their environment and interact with neighboring neural crest and other cell types. These mechanisms – contact inhibition of locomotion (CIL), co-attraction, and chemotaxis – function cooperatively to pattern the cranial NC into their migratory streams. In particular, the collective migration of the cranial NC depends on co-attraction and contact inhibition of locomotion (Carmona-Fontaine et al., 2008; Rogers et al., 2013; Scarpa et al., 2015). CIL, first described in an *in vitro* study of fibroblast cells, has been shown to be important for explaining neural crest migration (Abercrombie & Heaysman, 1953). This phenomenon is best described as when a cell ceases moving in the same direction upon contact with another cell and instead moves away in the opposite direction (Abercrombie, 1979). Mayor and colleagues, using live imaging in both *Xenopus* and zebrafish embryos, have shown the importance of CIL during the directional migration of CNC cells both *in vitro* and *in vivo* (Carmona-Fontaine et al. 2008, 2011). They found that the leading edge of *Xenopus* NC explants stopped their migration upon contact with NC cells from another NC explant, but were able to consume mesodermal explants, demonstrating that NC-NC interactions are homotypic. Live time-lapse microscopy has shown that NC cells collapse their lamellipodia upon contact with other NC cells, an observation similarly shown in fibroblast cells (Carmona-Fontaine et al. 2008, 2011). Thus, NC cells exposed at the leading edge of the explants migrate away from the cluster, leading to directional migration of the entire explant. Similarly, live imaging of zebrafish neural crest has demonstrated collapse of protrusions and changes in migratory direction when two NC cells make contact *in vivo*. Altogether, both *in vitro* and *in vivo* studies have shown that the directionality of NC migration is highly dependent on cell-cell interactions, from either another NC or other cell types, and CIL is required for normal neural crest migration.

The WNT/Planar Cell Polarity (PCP) pathway has been shown to mediates contact inhibition in *Xenopus* neural crest (Carmona-Fontaine et al., 2008; Mayor & Theveneau, 2014). Disruption of the PCP pathway results in failure of lamellipodial collapse when NC cells collide, resulting in maintenance of their direction of migration rather than CIL. N-cadherin plays a vital role in this process, mainly by inhibiting protrusion and *Rac1* activity at sites of cell contact while increasing *Rac1* activity at the leading edges. The polarized distribution of *Rac1* activity at the leading edge of the NC cell has also been shown to be crucial for responding to chemoattractants (Theveneau et al., 2010). In particular, the Stromal cell-derived factor 1 *(Sdf1/Cxcl12)* chemokine influences the migratory direction of NC cells *in vivo*, as demonstrated via loss-of-function experiments of either *Cxcl12* or its receptor *Cxcr4* in *Xenopus* embryos (Theveneau et al., 2010). *Cxcl12* binds to *Cxcr4* on the cell membrane and promotes intracellular *Rac1*, a key player in the activation and stabilization of the cell motility structures (e.g., filopodia, pseudopodia) that promote cell movement. Consistent with this, N-cadherin function-blocking antibodies cause a loss of attraction toward *Cxcl12*, emphasizing the importance of N-cadherin in maintaining functional cell–cell contacts during NC migration (Theveneau et al., 2010, 2013). *Sdf1/Cxcl12* interactions stabilize the large cell protrusions present at the leading edge through an increase in *Rac1* activity (Theveneau et al., 2010, 2013). Thus, *Cxcl12* may amplify and stabilize contact-dependent cell polarity to result in directional collective migration. Moreover, *Cxcl12* and its receptors *Cxcr4* and *Cxcr7* not only direct cranial neural crest cells into the branchial arches but also play a role in the condensation of the ventral cartilages in zebrafish (Olesnicky Killian et al., 2009).

Another factor important for neural crest chemotaxis and collective migration in *Xenopus* is the complement factor C3a and its receptor C3aR (Carmona-Fontaine et al., 2011; Theveneau et al., 2010). These act together with *Cxcl12/Cxcr4* to maintain co-attraction via cohesive clustering of migrating cranial crest cells. This phenomenon of co-attraction allows collective migration of neural crest cells despite low cell–cell adhesion and dispersion induced by CIL. As NC cells migrate, they may stray from the stream to which they belong. When this occurs, these cells are quickly attracted back to the stream via C3a-dependent chemotaxis which activates *Rac1*, polarizing the NC cells to return to the bulk of migratory stream.

Other cell types also influence collective cell migration. For example, the presence of cranial ectodermal placodes help shape the migratory behavior of cranial NC cells as they migrate through the developing *Xenopus* head. Collective cell migration occurs when NC cells "chase" placodal cells that in turn "run" upon contact with NC cells (Theveneau et al., 2010, 2013). When cultured separately *in vitro*, NC cells migrate randomly whereas placodes cells remain stationary. However, when co-cultured together, the placodal cells express the chemoattractant *Cxcl12* that attracts NC cells, resulting in the formation of transient, but functional, cell–cell adhesion complexes between NC and placode cells. Similar to CIL and dependent on N-cadherin PCP signaling, placode cell protrusions become destabilized upon physical contact with NC cells and in turn the placode cells "run" from the NC. Thus, the "chase-and-run" interaction between NC and placodal cells also promotes NC migration and patterning of the placode.

Another type of interaction between avian neural crest cells and the environment that results in the formation of segregated migratory "streams" is influenced by Ephrin-Eph signaling. These inhibitory signals parse NC cells into distinct streams such that they migrate through regions that lack the inhibitory Eph ligand (Gammill et al., 2007). For instance, neural crest cells expressing the tyrosine kinase receptor *EphA4+/ EphB1+* invade only BA3 and BA4, while neural crest cells expressing *Ephrin-B2* are repelled and migrate into BA2 (Osborne et al., 2005). In *Xenopus*, cranial neural crest cells do not form distinct streams, but do not intermix as they migrate, due to Eph/ephrin signaling (Smith et al., 1997). In the mouse, *Ephrin-B2* is expressed in the ectoderm of the branchial arches and neural epithelium. Its knock down results in failure of neural crest cells to migrate and populate BA2, even though the migratory streams remain segregated (Adams et al., 2001). On the other hand, knock-down of *Ephrin-B1* in cranial neural crest cells results in cells wandering between streams and migrating into BA3 and BA4 (Davy et al., 2004). It is not yet known which *Eph* receptors are involved or where they are expressed, but in situ hybridization in the chick suggests that complementary expression of *EphA* and *B* receptors on the cranial crest and *Ephrin-B1* and *EphB2* ligands on non-neural cells in the crest-free regions may play a role in segregating the streams thus promoting directional migration into the branchial arches (Mellott & Burke, 2008). Among other inhibitory signals, *Semaphorins*, *Plexins*, *Neuropilins*, and *Slit-Robo* interactions also influence NC migratory streams (reviewed in Minoux & Rijli, 2010).

Endothelin signaling also influences migration and final localization of NC cells. Many migratory cranial and cardiac NC cells express *Cxcr4* and endothelin receptors *EdnrA/B*. These cells are guided by neighboring tissues that express the cognate ligands, such as *Edn3* and *Sdf1/Cxcl12* (Square et al., 2020; Theveneau & Mayor, 2012). These ligand-receptor pairs have been shown to guide the migration and development of NC-derived to the heart and gut, respectively, to form the enteric nervous system (Ling & Sauka-Spengler, 2019; Square et al., 2020; Theveneau & Mayor, 2012). Additionally, other chemoattractants implicated to NC migration include VEGF, which is critical for CNC migration in the avian embryos (McLennan et al., 2015) and GDNF, critical for migration of vagal neural crest cells in populating the enteric nervous system (Sasselli et al., 2012). These signals, both attractive and inhibitory, help prevent dispersion and maintain the high density of the NC cells within their migration routes as they reach to their final destinations.

4.1 Trunk neural crest migration

In the trunk of the avian embryo, neural crest cells are thought to be specified by similar mechanisms as in the head but undergo different types of EMT and follow very different migratory pathways. Although avian cranial NC cells depart from the neural tube nearly all at once, the trunk NC delaminate progressively in a single chainlike manner after detaching from the neural tube (Cheung et al., 2005; Hall, 2008; Johnston, 1966; Le Douarin & Kalcheim, 1999; Le Douarin & Teillet, 1974; Noden, 1975; Théveneau et al., 2007). Upon closure of the neural tube, the trunk NC cells delaminate and migrate away along two main routes – either dorsolaterally under the ectoderm or ventromedially along the side of the neural tube (Fig. 1F). Interestingly, NC migration in the trunk is tightly linked with differentiation of the somites (Sela-Donenfeld & Kalcheim, 1999). Quail–chick chimeras have shown that the dorsolateral stream is primarily comprised of presumptive melanocytes. While the NC cells from the ventromedial stream mainly contribute to the sensory ganglia, sympathetic ganglia, adrenal medulla, aortic plexus, and some cells of the metanephric mesenchyme (George et al. 2007, 2010; Lefcort & George, 2007).

Initial studies focused on the migratory routes taken by the ventromedially migrating NC cells came from analyses using the HNK-1 antibody, which labels most migratory NC cells (Bronner-Fraser, 1986; Rickmann et al., 1985). These studies revealed that avian trunk NC cells invade the sclerotomes and migrate throughout rostral half of each somitic sclerotome, while being restricted from the caudal half, resulting in segmental migration. These observations propelled a search for possible rostrally localized attractants and/or caudally localized repellants. The caudal sclerotome expresses *Ephrins* and *Semaphorins*, along with inhibitory ECM molecules, such as *Versicans*, and *F-spondin* (Newgreen & Gooday, 1985; Newgreen et al., 1986). *Semaphorins* expressed in the caudal half of each sclerotome appear to be the primary reason for exclusion of neural crest cells from the caudal sclerotome by interacting with *Neuropilin2* receptors on neural crest cells, thus confining them to the rostral half of each somite which is repressor-free (Davy & Soriano, 2007; Gammill et al., 2007; Krull, 1998; Wang & Anderson, 1997). After migration, ventromedially migrating NC cells differentiate into dorsal root ganglia, with high levels of *Notch* promoting a glial fate, while high levels of *Delta* lead to a neuronal fate (Harris & Erickson, 2007; Krull et al., 1997; Krull, 1998; Wang & Anderson, 1997; Yu & Moens, 2005). Migration of these cells is also controlled temporally in a chemotactic fashion by the levels of *Cxcl12* along the ventro-dorsal axis, suggesting that a combination of attractive and repulsive cues guide trunk NC migration (Kasemeier-Kulesa et al., 2010).

The dorsolateral pathway, located between the future epidermis and dorsal somite/dermomyotome, is followed by avian neural crest cells destined to become melanocytes (Tosney, 2004). Repressors like *Slit1*, *Endothelin-3*, and *EphrinB1* are initially expressed under the epidermis on the dorsolateral route in the chick. Thus, early migrating trunk neural crest cells are initially excluded from this pathway. Later trunk NC cells, that express the receptor, *EphB2*, are able to migrate through these regions, as *Ephrins* are down-regulated. Cells moving along the dorsolateral pathway express *FoxD3* which induces MITF, a transcription factor necessary for the differentiation of neural crest cells into melanocytes (Thomas & Erickson, 2009).

Taken together, it is clear that various strategies guide neural crest cells along their migratory journey to their final destinations. The different mechanisms described above are under complex regulatory control that leads to intricate and stereotypic cell behaviors unique to each axial level. These not only guide neural crest cells to the proper locations but also are instrumental in ensuring appropriate differentiation. Despite many studies on neural crest migration, open questions remain about what direct these cells away from the neural tube and what signals cause cessation of migration and the transition to differentiation. While several signaling systems have been identified, there are others likely to be discovered in the future.

5. Onset of cranial neural crest ectomesenchymal differentiation

Following extensive migration throughout the developing embryo, guided by local environmental cues, neural crest cells progressively settle in diverse distal sites and utilize distinct gene regulatory networks to regulate their terminal differentiation. Although any given NC cell may have substantial flexibility to differentiate into various cell types, there is also evidence for lineage restrictions along the anteroposterior axis. The transition to a differentiated state is regulated by multiple transcription factors that co-bind to downstream loci to effectuate cell fate commitment and differentiation into a plethora of neural crest derivatives, ranging from chondrocytes, smooth muscle, peripheral neurons to glia, and melanocytes, among others (Le Douarin & Kalcheim, 1999; Le Douarin, 1982; reviewed in Martik & Bronner, 2017). Similar to NC induction, the gene regulatory networks for differentiation are highly dependent upon extracellular signaling cues. For example, chondrocyte differentiation is regulated by transforming growth factor (TGF-β) signaling (Furumatsu et al., 2005), while autonomic neurons are dependent on cues from BMP signaling (Morikawa et al., 2009) and melanocyte differentiation is regulated by WNT signaling (Jin et al., 2001; Takeda et al., 2000). Interestingly, transcription factors are often reused reiteratively to drive various developmental lineages. For instance, *Sox10*, plus *Mitf* and *Tfap2* regulate melanocyte differentiation in the periphery, while in the central nervous system, *Sox10* with *Olig1/2* mediate the differentiation of oligodendrocytes (Green et al., 2015; Potterf et al., 2001; Seberg et al., 2017; Simoes-Costa & Bronner, 2016; Soldatov et al., 2019).

These environmental cues and distinct gene batteries result in the activation or repression of effector genes, and in turn alter the transcriptional outputs of these cells (Simões-Costa & Bronner, 2015). Despite ample studies focused on how NC cells migrate to different regions of the embryo and how they are subjected to different spatial signaling cues, many open questions remain regarding what ultimately dictates their terminal fate commitment.

Classical lineage tracing approaches in chick have shown that the cranial neural crest contributes to cartilage and bone of the head, cranial nerves (CN), Schwann cells, the ossicles of the middle ear and periocular structures, smooth muscles, and connective tissues of the blood vessels. Additionally, these cells contribute to formation of the dermis of the head, teeth, melanocytes, and forebrain meninges (Fig. 2) (Couly et al., 1993; Evans & Noden, 2006; Gross & Hanken, 2005, 2008; Hall, 2008; Helms & Schneider, 2003; Le Douarin & Kalcheim, 1999; Le Lièvre, 1978; Minoux & Rijli, 2010; Moody & Heaton 1983c, 1983b, 1983a; Trainor & Tam, 1995). Finally, these cells also form multiple parts of the teeth, including dentin which is deposited and mineralized by NC-derived odontoblasts (Graveson et al., 1997; Lumsden & Buchanan, 1986; Lumsden, 1988; Smith & Hall, 1990). These derivatives are exclusive to the cranial NC population.

The cranial NC is thought to be a developmentally heterogeneous comprised of a combination of multipotent and partially fate-restricted cells (Cohen & Konigsberg, 1975; Dupin et al., 1990; Ito et al., 1993; Sieber-Blum & Cohen, 1980; Sieber-Blum, 1989). For example, lineage tracing approaches in zebrafish have revealed a spatial correspondence between the time of emigration and final fate of cranial NC cells (Schilling & Kimmel, 1994). Whereas early-migratory cranial NC cells give rise to skeletal and connective tissue, late-migrating cranial NC tend to assume neuronal fates. In both avian and mice embryos, lineage tracing and clonal neural crest cell differentiation studies have shown that these cells have both ectomesenchymal and neural potential, providing evidence that cranial neural crest cells are multipotent but become fate-restricted as they migrate towards their final destinations (Calloni et al., 2009). Consistent with these findings, cranial NC cells express *Sox10* and *FoxD3* (Britsch et al., 2001; Dottori et al., 2001) but these genes are down-regulated as the cells enter the branchial arches where expression of ectomesenchymal markers is upregulated.

Unlike the cranial neural crest, avian trunk NC cells do not form cartilage and are unable to differentiate into cartilage when grafted to the cranial region (Chibon, 1967; Le Douarin, 1982; Lwigale, 2001; Nakamura

& Lievre, 1982; Simoes-Costa & Bronner, 2016). The core regulatory network driving chondrocyte differentiation involves TGF-β signaling and the expression of *Sox5/6* and *Sox9*, which directly activate *Col2a1* and *Agc1*, both cartilage differentiation markers (Bell et al., 1997; Furumatsu et al., 2005; Ito et al., 2002; Lefebvre et al., 1998).

Taking advantage of this knowledge, Simões-Costa & Bronner utilized cranial versus trunk avian NC-specific enhancers to isolate pure populations of NC cells. Transcriptional profiling identified various transcription factors enriched in the cranial NC compared to the trunk NC: *Brn3c*, *Lhx5*, *Dmbx1*, *Ets1*, *Sox8*, and *Tfap2β* (Simoes-Costa & Bronner, 2016). Through a series of loss-of-function and promoter-chromatin immunoprecipitation (ChIP) assays, the regulatory linkage between these factors was established. It was revealed that ectopic expression of *Ets1*, *Sox8*, and *Tfap2β* in the trunk NC resulted in upregulation of chondrogenic genes, such as *Alx1* and *Runx2*, while concomitantly downregulating trunk NC genes, such as *Dbx2* and *Hes6*. Following reprogramming and transplantation into the cranial region, these cells gave rise to ectopic cartilage nodules. This was among the first demonstration that neural crest cells from other axial levels can be successfully reprogrammed into a cranial-like fate in avian embryos (Simoes-Costa & Bronner, 2016). Murine single cell profiling of cranial NC revealed that this population may transiently reactivate Yamanaka pluripotency factors (*Nanog*, *Oct4* and *Sox2*) before NC formation (Zalc et al., 2021). Thus, an *in vivo* programming event may be redirecting cranial NC towards multipotency via extensive epigenomic priming and chromatin remodeling and allowing these cells to generate ectomesenchymal derivatives and pattern the vertebrate head (Scerbo & Monsoro-Burq, 2020).

5.1 Cranial neural crest and placode contribution to peripheral nervous systems

Within the developing vertebrate head, the peripheral nervous systems (PNS) originates from two transient cell populations: the neural crest and ectodermal placode. Several cranial sensory ganglia are a mosaic of these two populations, organized together not only morphologically but also functionally as coherent structures. For example, both neural crest- and placode-derived cells contribute to distinct proximal and distal components, respectively, of facial (VII), glossopharyngeal (IX), and vagus (X) cranial nerves (Fig. 4) (Barlow, 2002; D'Amico-Martel & Noden, 1983; Steventon et al., 2014). In the case of the trigeminal ganglion, placode cells

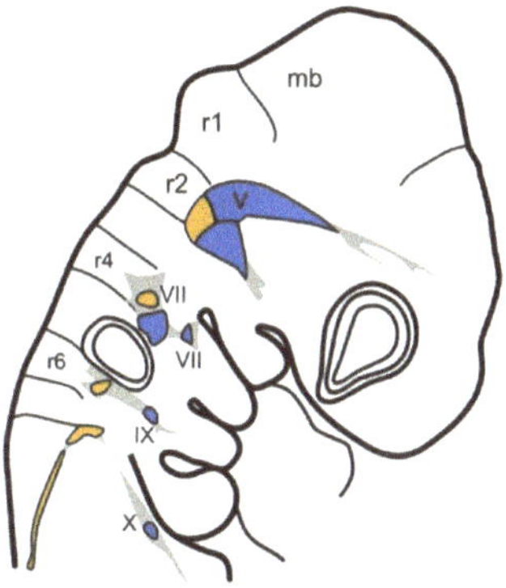

Fig. 4 Organization of cranial sensory ganglia. Schematic representing the ganglia and major nerve branches of the Vth, VIIth, VIIIth, IXth and Xth cranial nerves, as well as their embryonic origins (yellow, neural crest; blue, epibranchial placode). Neurons from the epibranchial, trigeminal, and otic placodes coalesce to form the distal aspects of the sensory ganglia (blue). Cranial NC cells form the proximal aspects (yellow), as well as ectome- senchymal cells, Schwann cells, melanocytes, and neurons.

give rise to sensory neurons mostly within the distal region, while neural crest cells generate neurons within the proximal region and glia throughout (Hamburger, 1961; Lwigale, 2001).

During early development, placodal cells thicken within the ectoderm, delaminate, and migrate into the cranial mesenchyme, after neural crest migration is well underway. In cases where both populations contribute to the same ganglia, placodal cells are in close proximity to migratory cranial NC cells, while other placodal populations, such as the otic, potentially act as barriers shaping the migratory streams of NC cells emerging from the hindbrain and heading towards the branchial arches (reviewed in Steventon et al., 2014). On the other hand, cranial neural crest cells may physically segregate and individualize placode-derived ganglionic clusters, which in turn enables formation of the NC streams within the head (Szabó et al., 2019; Theveneau et al., 2013). These types of intercellular interactions are also thought to be governed by the "chase-and-run" which aid in the proper migration and shaping of each cell population into ganglia (Szabó et al., 2019; Theveneau et al. 2010, 2013).

More recently, biomechanical studies have identified mechanical forces as 'trigger and guidance' cues for collective NC migration and differentiation (Barriga et al., 2018; Shellard & Mayor, 2019, 2021; Zhu et al., 2019). In particular, higher cell density of the cranial mesenchyme in *Xenopus* results in tissue stiffening, which neural crest cells detect. This in turn initiates collective cranial neural crest migration (Barriga et al., 2018). Furthermore, these studies have revealed that NC cells establish their migratory routes by detecting and following a stiffness gradient across adjacent tissues (Shellard & Mayor, 2019, 2021). In *Xenopus* embryos, this gradient is partly established by attractive signals from ventrally localized placode cells adjacent to cranial neural crest cells (Theveneau et al., 2013). Mechanical forces from placode cells also contribute to NC directional migration (Shellard & Mayor, 2021). Cranial NC cells 'soften' placode cells via N-cadherin interactions, thereby creating a collective durotaxis gradient that NC cells probe and follow during embryonic development. Ectopic stiffness gradients or absence of these gradients result in erratic NC migration *in vivo* (Shellard & Mayor, 2021). NC differentiation in mice is also affected by the rigidity of substrates. Stiff substrates result in smooth muscle differentiation whereas softer environments lead to glial differentiation (Li et al., 2011; Zhu et al., 2019). Taken together, these results demonstrate that mechanical cues drive collective migration and differentiation of NC cells. The molecular mechanisms driving the detection of these mechanical forces will be an interesting topic for further exploration.

Both modern and classical embryological experiments have demonstrated the interdependence of NC and placodal cell populations, both required for proper patterning of the cranial PNS (Bina et al., 2023; Gammill et al., 2007; Halmi et al., 2024; Hines & Taneyhill, 2024; Shiau & Bronner-Fraser, 2009; Shiau et al., 2008; Theveneau et al. 2010, 2013; Urrutia et al., 2024; Wu & Taneyhill, 2019). Ablation of the NC population does not lead to loss of cranial ganglia, but rather causes inappropriate morphology and positioning of ganglion, with evident abnormal or absent proximal components that lack projections to the CNS (Begbie & Graham, 2001; Cheng et al., 2000; Kuratani et al., 1991; Shiau et al., 2008; Yntema, 1944). Genetic ablation of the neural crest or perturbation of proper neural crest migration can lead to inappropriate fusions of otherwise physically separated ganglia in avian embryos (Bina et al., 2023; Gassmann et al., 1995; Golding et al., 2004; Hines & Taneyhill, 2024; Osborne et al., 2005; Schwarz et al., 2008; Shiau & Bronner-Fraser, 2009; Shiau et al., 2008; Urrutia et al., 2024; Wu & Taneyhill, 2019). On the other hand when placodes populations are ablated, NC cells are able to properly migrate but fail to produce cranial ganglia, with normal distal portions (Begbie & Graham, 2001; Cheng et al., 2000; Shiau et al., 2008). Finally, there is evidence that cranial NC cells actively guide placode cells by creating "corridors" that orchestrate the patterning of sensory neurons derived from both NC and placode cells (Freter et al., 2013). Altogether, these results point to the crucial role for NC and placode interactions during formation of cranial sensory ganglia.

5.2 Epibranchial ganglia

The epibranchial placodes contribute to sensory neurons in the distal portion of cranial ganglia VII, IXth and Xth (geniculate, petrosal and nodose placodes, respectively) (Fig. 4). These neurons innervate the taste buds of the oropharyngeal cavity, along with the heart, respiratory system, gastrointestinal tract and external ear. They relay vital sensory information via central projections to the hindbrain (D'Amico-Martel & Noden, 1983; Harlow et al., 2011; Harlow & Barlow, 2007; Schlosser & Northcutt, 2000). Lineage labeling with CM-DiI and CM-DiO of NC and placode cells, respectively, has revealed that placode-derived neurons overlap with neurogliogenic neural crest cells (Begbie & Graham, 2001). In addition to placodally derived neurons, NC cells contribute to proximal region of these ganglia, forming both peripheral glial cells (Schwann cells) and sensory neurons (D'Amico-Martel & Noden, 1983; Harlow & Barlow, 2007; Harlow et al., 2011; Le Douarin et al., 1991; Le Douarin, 1986; Schoenwolf et al., 2009).

Classical embryological studies have provided a clear understanding of the cellular contributions of neural crest and placode cells in the formation of epibranchial ganglia. In particular, ablation of prospective geniculate ectoderm in the chick results in ventrolateral displacement and loss or reduction of the central connections to the CNS (Yntema, 1944). Similarly, ablation of the cardiac neural crest results in similar phenotypes in the petrosal and nodose ganglia. In these instances, the distal portion of the ganglia develops in complete isolation, but the proximal and central regions are entirely lost. Later studies also confirmed these observations in the chick (Begbie & Graham, 2001; Kuratani et al., 1991; Yntema, 1944). In addition, placode-derived neurons fail to internally migrate resulting in displaced ganglia (Begbie & Graham, 2001). Thus, neural crest cells are required for accurate positioning of the epibranchial ganglia, as well as for establishing appropriate connections with the CNS.

In developing mouse embryos, proneural genes *Neurog1* and *Neurog2* are transiently expressed in distinct cranial ganglia. In particular, the geniculate and petrosal ganglia express *Neurog2*, while the nodose ganglia express both *Neurog1* and *Neurog2* (Fode et al., 1998; Ma et al., 1998). In *Neurog2* null mutant mice, the delamination and differentiation of the geniculate and petrosal placode progenitors are altered, resulting in delayed development of the distal portion of the geniculate ganglia and loss of the petrosal ganglia, mainly due to epibranchial neuroblasts failing to delaminate (Coppola et al., 2010). While the nodose ganglion remains unaffected with single mutants, *Neurog1* and *Neurog2* double mutants result in loss of the ganglion, suggesting that there may function redundantly (Espinosa-Medina et al., 2014; Fode et al., 1998; Ma et al., 1998; Takano-Maruyama et al., 2012). Thus, *Neurog1/2* are proneural factors critical for initiating and balancing the neuronal differentiation of progenitors within the epibranchial ganglia (Bertrand et al., 2002; Huang et al., 2014). Coppola and colleagues (2010) observed not only a loss of the geniculate ganglia in *Ngn2* knock-out mice, but also of the sphenopalatine sympathetic ganglion, raising the possibility that placodal neurons provide a survival signal for NC-derived neurons which they innervate. Together, these studies further point to reciprocal interaction between neural crest and placodes cells in the establishment of correct circuits within the peripheral nervous system.

5.3 Trigeminal ganglia

The reciprocal interactions between NC and placode cells have also been investigated during formation of the trigeminal ganglion. Unlike the epibranchial ganglia, which have clear NC and placodal contribution to the proximal and distal regions, respectively, the trigeminal ganglion is comprised of a mixture of neurons derived from both embryonic populations (Baker, 2006; Blentic et al., 2011; D'Amico-Martel, 1982; D'Amico-Martel & Noden, 1983; Hamburger, 1961; Jidigam & Gunhaga, 2013; Knecht & Bronner-Fraser, 2002; Lwigale, 2001; Saint-Jeannet & Moody, 2014; Shiau et al., 2008). The largest of the cranial ganglia, the trigeminal ganglion contains sensory neurons derived from both NC and placode cell, with the placode (the maxillomandibular and ophthalmic placodes in amniotes) contributing predominantly to the distal regions of the ganglion, while neural crest cells contribute proximally, along with associated glia throughout in developing birds (D'Amico-Martel & Noden, 1983; Hamburger, 1961; Moody & Heaton, 1983a, 1983b, 1983c; Moody et al., 1989; Noden, 1975; Serbedzija et al., 1992; Shiau et al., 2008). Interestingly, within both the epibranchial and trigeminal ganglia, terminal differentiation of placodally-derived neurons precedes that of NC-derived neurons. As a result, cranial neural crest cells do not undergo large-scale neuronal differentiation until after the trigeminal ganglion has fully condensed (Begbie et al., 2002; d'Amico-Martel & Noden, 1980; D'Amico-

Martel & Noden, 1983; Hamburger, 1961; Moody & Heaton, 1983a, 1983b, 1983c; Moody et al., 1989; Noden, 1975; Serbedzija et al., 1992; Shiau et al., 2008). To determine if whether NC cells play a role in the induction of ophthalmic placode cells, ablation studies in chick, followed by analysis of *Pax3* expression, revealed that loss of NC cells did not change *Pax3* expression but instead resulted in ganglia that were abnormal and misshapen (D'Amico-Martel & Noden, 1983; Gammill et al., 2007; Hamburger, 1961; Shiau et al., 2008; Stark et al., 1997). Later NC ablations result in formation of slightly reduced, but overall normal ganglia. However, condensation of the ganglia was affected, along with a lack of axon projections to the hindbrain (Hamburger, 1961; Moody & Heaton, 1983c). Disruption of NC migration due to changes in *Neuropilin/Semaphorin* signaling or loss of cadherin molecules also disrupt trigeminal condensation and gangliogenesis, further supporting a role for NC cells in facilitating the aggregation of placodal neurons into the trigeminal ganglion (Akitaya & Bronner-Fraser, 1992; D'Amico-Martel & Norden, 1983; Gammill et al., 2007; Saint-Jeannet & Moody, 2014; Shiau & Bronner-Fraser, 2009; Shiau et al., 2008). Together, these results illustrate the vital role of reciprocal interactions between NC and placode cells in ganglion condensation and establishment of appropriate central connections.

At a molecular level, various adhesion and signaling molecules have been shown to facilitate trigeminal ganglion formation. In avian embryos, *Slit-Robo* interactions are one of these molecular mechanism by which NC and placode interactions aid in the condensation and formation of the ganglia (Shiau & Bronner-Fraser, 2009; Shiau et al., 2008). Loss-of-function experiments have shown that loss of either component results in malformation of the ganglion (Shiau et al., 2008). It was later determined that N-cadherin may be acting downstream of *Slit-Robo*, aiding in the aggregation of the trigeminal ganglion (Shiau & Bronner-Fraser, 2009). Electroporation of morpholinos against N-cadherin into the presumptive trigeminal ectoderm, leads to dispersed and mislocalized trigeminal neuronal projections; however, ingression of these placodally-derived neurons remained unaffected (Shiau & Bronner-Fraser, 2009). Interestingly, it was confirmed that N-cadherin expression is downstream of *Robo2*, as *Slit1* overexpression results in ectopic aggregates of N-cadherin positive and *Robo2* negative neurons (Shiau & Bronner-Fraser, 2009).

In addition, previous studies have also shown that intercellular interactions during trigeminal ganglion formation are mediated, in part, by cadherin-based adhesion. In particular, Cadherin-7 and N-cadherin are expressed in neural crest and placode cells, respectively, during formation of the chick trigeminal ganglia (Hatta et al., 1987; Nakagawa & Takeichi, 1995; Shiau & Bronner-Fraser, 2009; Wu & Taneyhill, 2019). Neither depletion nor ectopic expression of Cadherin-7 impacts the overall morphology of the ganglion. While knockdown of N-cadherin does not affect initial placode cell ingression and delamination, it leads to increased placodal neuron dispersal during trigeminal gangliogenesis (Wu & Taneyhill, 2019; Wu et al., 2014). Overall, the interactions between Cadherin-7, expressed in NC cells, and N-cadherin, present in placodal neurons, is required for coalescing of NC and placodal neurons during trigeminal gangliogenesis.

Additional signaling molecules have been implicated in the role of NC and placode interactions during cranial ganglion formation. In particular, disruption of *Neuropilin/Semaphorin* signaling affect cranial NC migratory behaviors (Osborne et al., 2005; Schwarz et al., 2008). In mice lacking the *Neuropilin* receptors NRP1/2, NC cells invade NC-free territories between the mandibular and hyoid NC streams, resulting in fused trigeminal and facial ganglia (Schwarz et al., 2008). Additionally, inhibition of the *Semaphorins Sema3A/3F* results in loss of NC migration and inhibits placodal aggregation in both the epibranchial and trigeminal ganglion (Osborne et al., 2005). Similarly, loss of the tyrosine kinase receptor, *erbB4* in either mice or chick embryos results in misrouting of r4 NC cells and production of ectopic cranial nerves (Gassmann et al., 1995; Golding et al., 2004).

Another signaling molecule implicated in promoting NC and placode condensation during chick trigeminal gangliogenesis is WNT6, along with its WNT modulator, *Wise* (Beaudoin et al., 2005; Guidato & Itasaki, 2007; Shigetani & Itasaki, 2007). Previous studies have shown that ectopic expression of *Wise* results in the formation of ectopic ganglia, a phenotype similar to the ectopic expression of *Wnt6*, suggesting that both *Wise* and *Wnt6* may act synergistically and thus may be inducing both placode and NC earlier than intended, resulting in these ectopic ganglia (Bastidas et al., 2004; Canning et al., 2008; García-Castro et al., 2002; Lassiter et al., 2007; Lewis et al., 2004; Shigetani & Itasaki, 2007).

In addition to adhesion and signaling molecules, multiple transcription factors have been implicated in promoting the condensation and differentiation of neurons with the trigeminal ganglion. Vertebrate neurogenesis is governed, in part, by proneural genes that regulate cell type determination and terminal differentiation. Basic Helix-Loop-Helix (bHLH) transcription factors, *Neurog1/2*, along with downstream transcription factors, such as

NeuroD1, have been identified in differentiating post-mitotic neuronal precursors and contribute to the establishment or maintenance of the neuronal program (Bertrand et al., 2002; Bina et al., 2023; Ma et al. 1996, 1998; Sommer et al., 1996; Ventéo et al., 2019). Depletion of *Neurog2* or *NeuroD1* from chick trigeminal placode cells results in malformation of the trigeminal ganglion. In particular, *Neurog2* and *NeuroD1* are shown to be critical for the development of the trigeminal ganglion, controlling neuronal differentiation, and precise axon outgrowth and innervation of target tissues as well as neuron morphology (Bina et al., 2023).

Another transcription factor implicated in differentiation and fate specification of sensory neurons in the peripheral nervous systems is *Tlx3* (Cheng et al., 2004; Hornbruch et al., 2005; Kondo et al., 2008; Logan et al., 1998; Xu et al., 2008). In the chick, previous studies have shown that *Tlx3* is expressed in placodally derived components of the trigeminal ganglion, serving as a post-mitotic selector gene propelling these cells towards a neuronal fate (Cheng et al., 2004; Kondo et al., 2008; Logan et al., 1998). In addition, recent studies have revealed that, in the chick, *Tlx3* is expressed by both NC cells and placode cells and required for proper gangliogenesis. Moreover, ectopic expression of *Tlx3* leads to premature expression of neural markers, while its depletion results in improper condensation and size reduction of the ganglion. This suggests that *Tlx3* is necessary for specification of neural crest- and placode-derived sensory neurons during development of the trigeminal ganglion (Urrutia et al., 2024).

Taken together, neural crest cells play a vital role in the correct aggregation and condensation to form the trigeminal ganglion and establish correct connections to the hindbrain. Importantly, these results suggest a dual role for the NC in guiding the initial migration of placodal neurons to the appropriate sites as well as contributing to neurons and glia within the trigeminal ganglion.

6. Concluding remarks

In this review, we attempt to summarize our current knowledge of neural crest formation, migratory behaviors, and contribution to cranial sensory gangliogenesis. The neural crest is a captivating embryonic population, due to its migratory ability, plethora of derivatives, and broad developmental potential. This unique cell population is responsible for formation of craniofacial cartilage and bones and the peripheral nervous system, both defining features of vertebrates, effectively parting them from invertebrate chordates. Through a network of transcription factors and signaling cascades, different events in neural crest development are tightly regulated. Regardless of axial level of origin, many neural crest cells are multipotent and are capable of differentiating into a diverse array of derivatives (Baggiolini et al., 2015). Our current understanding of the NC and its transcriptional regulation has revealed multiple layers of control. While it is clear that there are both direct and indirect interactions between different neural crest transcription factors, adhesion, and signaling molecules, our grasp of the precise regulation of normal NC development and differentiation, and how their dysregulation can contribute to disease remains a mystery.

The use of high-throughput sequencing together with classical perturbation techniques has greatly enhanced our knowledge of how the neural crest develops. Additionally, technologies such as ATAC-seq, ChIP-seq, single cell RNA-seq, CRISPR/Cas9, and high-resolution microscopy have opened the possibility to dig deeper into some of the fundamental questions in neural crest biology, for example, better detailing the contribution of neural crest and placode interactions to morphogenetic events. These new methodologies hold the promise of providing further insights into the molecular mechanisms that control complex tissue rearrangements during vertebrate development. These techniques also can be applied to understand the cues involved in assembling ganglia in correct positions, as well as overall morphogenesis of other neural crest-derived structures and how their dysregulation can cause tissue deformities. Finally, we expect that future work will continue expanding our understanding of how the assembly of peripheral structures is coordinated with their central targets. Recent studies suggest that some of the early mechanisms that pattern the CNS are equally involved in patterning the periphery, and vice versa.

Acknowledgments

We would like to thank Roberto Mayor for organizing this book volume. This work was supported by NIH grants R01DE027568 and R01DE024157 to M.E.B. and F31DE031154 to H.A.U.

References

Abercrombie, M. (1979). Contact inhibition and malignancy. *Nature, 281*, 259–262. https://doi.org/10.1038/281259a0.

Abercrombie, M., & Heaysman, J. E. (1953). Observations on the social behaviour of cells in tissue culture. I. Speed of movement of chick heart fibroblasts in relation to their mutual contacts. *Experimental Cell Research, 5*, 111–131. https://doi.org/10.1016/0014-4827(53)90098-6.

Adams, R. H., Diella, F., Hennig, S., Helmbacher, F., Deutsch, U., & Klein, R. (2001). The cytoplasmic domain of the ligand ephrinB2 is required for vascular morphogenesis but not cranial neural crest migration. *Cell, 104*, 57–69. https://doi.org/10.1016/s0092-8674(01)00191-x.

Akitaya, T., & Bronner-Fraser, M. (1992). Expression of cell adhesion molecules during initiation and cessation of neural crest cell migration. *Developmental Dynamics: an Official Publication of the American Association of Anatomists, 194*, 12–20. https://doi.org/10.1002/aja.1001940103.

Alfandari, D., Cousin, H., & Marsden, M. (2010). Mechanism of Xenopus cranial neural crest cell migration. *Cell Adhesion Migration, 4*, 553–560. https://doi.org/10.4161/cam.4.4.12202.

Baggiolini, A., Varum, S., Mateos, J. M., Bettosini, D., John, N., Bonalli, M., ... Sommer, L. (2015). Premigratory and migratory neural crest cells are multipotent in vivo. *Cell Stem Cell, 16*, 314–322. https://doi.org/10.1016/j.stem.2015.02.017.

Bajpai, R., Chen, D. A., Rada-Iglesias, A., Zhang, J., Xiong, Y., Helms, J., ... Wysocka, J. (2010). CHD7 cooperates with PBAF to control multipotent neural crest formation. *Nature, 463*, 958–962. https://doi.org/10.1038/nature08733.

Baker, C. (2006). *Neural crest and cranial ectodermal placodes. Developmental neurobiology*. New York: Kluwer Academic Publishers-Plenum Publishers 67–127. https://doi.org/10.1007/0-387-28117-7_4.

Baker, C. V. H. (2008). The evolution and elaboration of vertebrate neural crest cells. *Current Opinion in Genetics & Development, 18*, 536–543. https://doi.org/10.1016/j.gde.2008.11.006.

Barlow, L. A. (2002). Cranial nerve development: Placodal neurons ride the crest. *Current Biology: CB, 12*, R171–R173. https://doi.org/10.1016/s0960-9822(02)00734-0.

Barriga, E. H., Franze, K., Charras, G., & Mayor, R. (2018). Tissue stiffening coordinates morphogenesis by triggering collective cell migration in vivo. *Nature, 554*, 523–527. https://doi.org/10.1038/nature25742.

Barriga, E. H., Maxwell, P. H., Reyes, A. E., & Mayor, R. (2013). The hypoxia factor Hif-1α controls neural crest chemotaxis and epithelial to mesenchymal transition. *The Journal of Cell Biology, 201*, 759–776. https://doi.org/10.1083/jcb.201212100.

Basch, M. L., Bronner-Fraser, M., & García-Castro, M. I. (2006). Specification of the neural crest occurs during gastrulation and requires Pax7. *Nature, 441*, 218–222. https://doi.org/10.1038/nature04684.

Bastidas, F., De Calisto, J., & Mayor, R. (2004). Identification of neural crest competence territory: Role of Wnt signaling. *Developmental Dynamics: an Official Publication of the American Association of Anatomists, 229*, 109–117. https://doi.org/10.1002/dvdy.10486.

Beaudoin, G. M. J., 3rd, Sisk, J. M., Coulombe, P. A., & Thompson, C. C. (2005). Hairless triggers reactivation of hair growth by promoting Wnt signaling. *Proceedings of the National Academy of Sciences of the United States of America, 102*, 14653–14658. https://doi.org/10.1073/pnas.0507609102.

Begbie, J., Ballivet, M., & Graham, A. (2002). Early steps in the production of sensory neurons by the neurogenic placodes. *Molecular and Cellular Neurosciences, 21*, 502–511. https://doi.org/10.1006/mcne.2002.1197.

Begbie, J., & Graham, A. (2001). Integration between the epibranchial placodes and the hindbrain. *Science, 294*, 595–598. https://doi.org/10.1126/science.1062028.

Bell, D. M., Leung, K. K., Wheatley, S. C., Ng, L. J., Zhou, S., Ling, K. W., ... Cheah, K. S. (1997). SOX9 directly regulates the type-II collagen gene. *Nature Genetics, 16*, 174–178. https://doi.org/10.1038/ng0697-174.

Bertrand, N., Castro, D. S., & Guillemot, F. (2002). Proneural genes and the specification of neural cell types. *Nature Reviews. Neuroscience, 3*, 517–530. https://doi.org/10.1038/nrn874.

Betancur, P., Bronner-Fraser, M., & Sauka-Spengler, T. (2010). Assembling neural crest regulatory circuits into a gene regulatory network. *Annual Review of Cell and Developmental Biology, 26*, 581–603. https://doi.org/10.1146/annurev.cellbio.042308.113245.

Bhattacharyya, S., Bailey, A. P., Bronner-Fraser, M., & Streit, A. (2004). Segregation of lens and olfactory precursors from a common territory: Cell sorting and reciprocity of Dlx5 and Pax6 expression. *Developmental Biology, 271*, 403–414. https://doi.org/10.1016/j.ydbio.2004.04.010.

Bina, P., Hines, M. A., Sanyal, J., & Taneyhill, L. A. (2023). Neurogenin 2 and neuronal differentiation 1 control proper development of the chick trigeminal ganglion and its nerve branches. *Journal of Developmental Biology, 11*. https://doi.org/10.3390/jdb11010008.

Blentic, A., Chambers, D., Skinner, A., Begbie, J., & Graham, A. (2011). The formation of the cranial ganglia by placodally-derived sensory neuronal precursors. *Molecular and Cellular Neurosciences, 46*, 452–459. https://doi.org/10.1016/j.mcn.2010.11.010.

Britsch, S., Goerich, D. E., Riethmacher, D., Peirano, R. I., Rossner, M., Nave, K. A., ... Wegner, M. (2001). The transcription factor Sox10 is a key regulator of peripheral glial development. *Genes & Development, 15*, 66–78. https://doi.org/10.1101/gad.186601.

Bronner, M. E., & LeDouarin, N. M. (2012). Development and evolution of the neural crest: An overview. *Developmental Biology, 366*, 2–9. https://doi.org/10.1016/j.ydbio.2011.12.042.

Bronner-Fraser, M. (1995). Origin of the avian neural crest. *Stem Cells, 13*, 640–646. https://doi.org/10.1002/stem.5530130608.

Bronner-Fraser, M. (1993). Neural crest cell migration in the developing embryo. *Trends in Cell Biology, 3*, 392–397. https://doi.org/10.1016/0962-8924(93)90089-j.

Bronner-Fraser, M. (1986). Analysis of the early stages of trunk neural crest migration in avian embryos using monoclonal antibody HNK-1. *Developmental Biology, 115*, 44–55. https://doi.org/10.1016/0012-1606(86)90226-5.

Calloni, G. W., Le Douarin, N. M., & Dupin, E. (2009). High frequency of cephalic neural crest cells shows coexistence of neurogenic, melanogenic, and osteogenic differentiation capacities. *Proceedings of the National Academy of Sciences of the United States of America, 106*, 8947–8952. https://doi.org/10.1073/pnas.0903780106.

Canning, C. A., Lee, L., Luo, S. X., Graham, A., & Jones, C. M. (2008). Neural tube derived Wnt signals cooperate with FGF signaling in the formation and differentiation of the trigeminal placodes. *Neural Development, 3*, 35. https://doi.org/10.1186/1749-8104-3-35.

Carmona-Fontaine, C., Matthews, H. K., Kuriyama, S., Moreno, M., Dunn, G. A., Parsons, M., ... Mayor, R. (2008). Contact inhibition of locomotion in vivo controls neural crest directional migration. *Nature, 456*, 957–961. https://doi.org/10.1038/nature07441.

Carmona-Fontaine, C., Theveneau, E., Tzekou, A., Tada, M., Woods, M., Page, K. M., ... Mayor, R. (2011). Complement fragment C3a controls mutual cell attraction during collective cell migration. *Developmental Cell, 21*, 1026–1037. https://doi.org/10.1016/j.devcel.2011.10.012.

Chalpe, A. J., Prasad, M., Henke, A. J., & Paulson, A. F. (2010). Regulation of cadherin expression in the chicken neural crest by the Wnt/β-catenin signaling pathway. *Cell Adhesion Migration, 4*, 431–438. https://doi.org/10.4161/cam.4.3.12138.

Cheng, L., Arata, A., Mizuguchi, R., Qian, Y., Karunaratne, A., Gray, P. A., ... Ma, Q. (2004). Tlx3 and Tlx1 are post-mitotic selector genes determining glutamatergic over GABAergic cell fates. *Nature Neuroscience*. https://doi.org/10.1038/nn1221.

Cheng, Y., Cheung, M., Abu-Elmagd, M. M., Orme, A., & Scotting, P. J. (2000). Chick sox10, a transcription factor expressed in both early neural crest cells and central nervous system. *Brain, 121*, 233–241. https://doi.org/10.1016/s0165-3806(00)00049-3.

Cheung, M., Chaboissier, M.-C., Mynett, A., Hirst, E., Schedl, A., & Briscoe, J. (2005). The transcriptional control of trunk neural crest induction, survival, and delamination. *Developmental Cell, 8*, 179–192. https://doi.org/10.1016/j.devcel.2004.12.010.

Chibon, P. (1967). Nuclear labelling by tritiated thymidine of neural crest derivatives in the amphibian Urodele Pleurodeles waltlii Michah. *Journal of Embryology and Experimental Morphology, 18*, 343–358.

Cohen, A. M., & Konigsberg, I. R. (1975). A clonal approach to the problem of neural crest determination. *Developmental Biology, 46*, 262–280. https://doi.org/10.1016/0012-1606(75)90104-9.

Coppola, E., Rallu, M., Richard, J., Dufour, S., Riethmacher, D., Guillemot, F., ... Brunet, J.-F. (2010). Epibranchial ganglia orchestrate the development of the cranial neurogenic crest. *The Proceedings of the National Academy of Sciences U. S. A. 107*, 2066–2071.

Couly, G. F., Coltey, P. M., & Le Douarin, N. M. (1993). The triple origin of skull in higher vertebrates: A study in quail-chick chimeras. *Development, 117*, 409–429. https://doi.org/10.1242/dev.117.2.409.

Couly, G. F., & Le Douarin, N. M. (1987). Mapping of the early neural primordium in quail-chick chimeras. II. The prosencephalic neural plate and neural folds: Implications for the genesis of cephalic human congenital abnormalities. *Developmental Biology, 120*, 198–214. https://doi.org/10.1016/0012-1606(87)90118-7.

Creuzet, S., Couly, G., Vincent, C., & Le Douarin, N. M. (2002). Negative effect of Hox gene expression on the development of the neural crest-derived facial skeleton. *Development, 129*, 4301–4313. https://doi.org/10.1242/dev.129.18.4301.

Dady, A., Blavet, C., & Duband, J.-L. (2012). Timing and kinetics of E- to N-cadherin switch during neurulation in the avian embryo. *Developmental Dynamics: an Official Publication of the American Association of Anatomists, 241*, 1333–1349. https://doi.org/10.1002/dvdy.23813.

D'Amico-Martel, A., & Noden, D. M. (1980). An autoradiographic analysis of the development of the chick trigeminal ganglion. *Journal of Embryology and Experimental Morphology, 55*, 167–182.

D'Amico-Martel, A. (1982). Temporal patterns of neurogenesis in avian cranial sensory and autonomic ganglia. *The American Journal of Anatomy, 163*, 351–372. https://doi.org/10.1002/aja.1001630407.

D'Amico-Martel, A., & Noden, D. M. (1983). Contributions of placodal and neural crest cells to avian cranial peripheral ganglia. *The American Journal of Anatomy, 166*, 445–468. https://doi.org/10.1002/aja.1001660406.

Davis, E. M., & Trinkaus, J. P. (1981). Significance of cell-to-cell contacts for the directional movement of neural crest cells within a hydrated collagen lattice. *Development, 63*, 29–51. https://doi.org/10.1242/dev.63.1.29.

Davy, A., Aubin, J., & Soriano, P. (2004). Ephrin-B1 forward and reverse signaling are required during mouse development. *Genes & Development, 18*, 572–583. https://doi.org/10.1101/gad.1171704.

Davy, A., & Soriano, P. (2007). Ephrin-B2 forward signaling regulates somite patterning and neural crest cell development. *Developmental Biology, 304*, 182–193. https://doi.org/10.1016/j.ydbio.2006.12.028.

Del Barrio, M. G., & Nieto, M. A. (2002). Overexpression of Snail family members highlights their ability to promote chick neural crest formation. *Development, 129*, 1583–1593. https://doi.org/10.1242/dev.129.7.1583.

Desai, R. A., Gao, L., Raghavan, S., Liu, W. F., & Chen, C. S. (2009). Cell polarity triggered by cell-cell adhesion via E-cadherin. *Journal of Cell Science, 122*, 905–911. https://doi.org/10.1242/jcs.028183.

Dottori, M., Gross, M. K., Labosky, P., & Goulding, M. (2001). The winged-helix transcription factor Foxd3 suppresses interneuron differentiation and promotes neural crest cell fate. *Development, 128*, 4127–4138. https://doi.org/10.1242/dev.128.21.4127.

Duband, J. L., & Thiery, J. P. (1982). Distribution of fibronectin in the early phase of avian cephalic neural crest cell migration. *Developmental Biology, 93*, 308–323. https://doi.org/10.1016/0012-1606(82)90120-8.

Dupin, E., Baroffio, A., Dulac, C., Cameron-Curry, P., & Le Douarin, N. M. (1990). Schwann-cell differentiation in clonal cultures of the neural crest, as evidenced by the anti-Schwann cell myelin protein monoclonal antibody. *Proceedings of the National Academy of Sciences of the United States of America, 87*, 1119–1123. https://doi.org/10.1073/pnas.87.3.1119.

Dupin, E., Creuzet, S., & Le Douarin, N. M. (2006). The contribution of the neural crest to the vertebrate body. *Advances in Experimental Medicine and Biology, 589*, 96–119. https://doi.org/10.1007/978-0-387-46954-6_6.

Dupin, I., Camand, E., & Etienne-Manneville, S. (2009). Classical cadherins control nucleus and centrosome position and cell polarity. *The Journal of Cell Biology, 185*, 779–786. https://doi.org/10.1083/jcb.200812034.

Erickson, C. A. (1985). Control of neural crest cell dispersion in the trunk of the avian embryo. *Developmental Biology, 111*, 138–157. https://doi.org/10.1016/0012-1606(85)90442-7.

Erickson, C. A., & Reedy, M. V. (1998). Neural crest development: The interplay between morphogenesis and cell differentiation. *Current Topics in Developmental Biology, 40*, 177–209. https://doi.org/10.1016/s0070-2153(08)60367-1.

Erickson, C. A., Tosney, K. W., & Weston, J. A. (1980). Analysis of migratory behavior of neural crest and fibroblastic cells in embryonic tissues. *Developmental Biology, 77*, 142–156. https://doi.org/10.1016/0012-1606(80)90462-5.

Espinosa-Medina, I., Outin, E., Picard, C. A., Chettouh, Z., Dymecki, S., Consalez, G. G., ... Brunet, J.-F. (2014). Neurodevelopment. Parasympathetic ganglia derive from Schwann cell precursors. *Science, 345*, 87–90. https://doi.org/10.1126/science.1253286.

Etienne-Manneville, S. (2014). Neighborly relations during collective migration. *Current Opinion in Cell Biology, 30*, 51–59. https://doi.org/10.1016/j.ceb.2014.06.004.

Evans, D. J. R., & Noden, D. M. (2006). Spatial relations between avian craniofacial neural crest and paraxial mesoderm cells. *Developmental Dynamics: an Official Publication of the American Association of Anatomists, 235*, 1310–1325. https://doi.org/10.1002/dvdy.20663.

Fairchild, C. L., Conway, J. P., Schiffmacher, A. T., Taneyhill, L. A., & Gammill, L. S. (2014). FoxD3 regulates cranial neural crest EMT via downregulation of tetraspanin18 independent of its functions during neural crest formation. *Mechanisms of Development, 132*, 1–12. https://doi.org/10.1016/j.mod.2014.02.004.

Fairchild, C. L., & Gammill, L. S. (2013). Tetraspanin18 is a FoxD3-responsive antagonist of cranial neural crest epithelial-to-mesenchymal transition that maintains cadherin-6B protein. *Journal of Cell Science, 126*, 1464–1476. https://doi.org/10.1242/jcs.120915.

Farlie, P. G., Kerr, R., Thomas, P., Symes, T., Minichiello, J., Hearn, C. J., & Newgreen, D. (1999). A paraxial exclusion zone creates patterned cranial neural crest cell outgrowth adjacent to rhombomeres 3 and 5. *Developmental Biology, 213*, 70–84. https://doi.org/10.1006/dbio.1999.9332.

Faulkner-Jones, B. E., Godinho, L. N., Reese, B. E., Pasquini, G. F., Ruefli, A., & Tan, S. S. (1999). Cloning and expression of mouse Cadherin-7, a type-II cadherin isolated from the developing eye. *Molecular and Cellular Neurosciences, 14*, 1–16. https://doi.org/10.1006/mcne.1999.0764.

Fode, C., Gradwohl, G., Morin, X., Dierich, A., LeMeur, M., Goridis, C., & Guillemot, F. (1998). The bHLH protein NEUROGENIN 2 is a determination factor for epibranchial placode-derived sensory neurons. *Neuron, 20*, 483–494. https://doi.org/10.1016/s0896-6273(00)80989-7.

Freter, S., Fleenor, S. J., Freter, R., Liu, K. J., & Begbie, J. (2013). Cranial neural crest cells form corridors prefiguring sensory neuroblast migration. *Development, 140*, 3595–3600. https://doi.org/10.1242/dev.091033.
Friedl, P., & Gilmour, D. (2009). Collective cell migration in morphogenesis, regeneration and cancer. *Nature Reviews. Molecular Cell Biology, 10*, 445–457. https://doi.org/10.1038/nrm2720.
Frith, T., Jr, Granata, I., Wind, M., Stout, E., Thompson, O., ... Tsakiridis, A. (2018). Human axial progenitors generate trunk neural crest cells in vitro. *Elife, 7*. https://doi.org/10.7554/eLife.35786.
Furumatsu, T., Tsuda, M., Taniguchi, N., Tajima, Y., & Asahara, H. (2005). Smad3 induces chondrogenesis through the activation of SOX9 via CREB-binding protein/p300 recruitment. *The Journal of Biological Chemistry, 280*, 8343–8350. https://doi.org/10.1074/jbc.M413913200.
Gammill, L. S., Gonzalez, C., & Bronner-Fraser, M. (2007). Neuropilin 2/semaphorin 3F signaling is essential for cranial neural crest migration and trigeminal ganglion condensation. *Developmental Neurobiology, 67*, 47–56. https://doi.org/10.1002/dneu.20326.
Gans, C., & Northcutt, R. G. (1983). Neural crest and the origin of vertebrates: A new head. *Science, 220*, 268–273. https://doi.org/10.1126/science.220.4594.268.
García-Castro, M. I., Marcelle, C., & Bronner-Fraser, M. (2002). Ectodermal Wnt function as a neural crest inducer. *Science, 297*, 848–851. https://doi.org/10.1126/science.1070824.
Gassmann, M., Casagranda, F., Orioli, D., Simon, H., Lai, C., Klein, R., & Lemke, G. (1995). Aberrant neural and cardiac development in mice lacking the ErbB4 neuregulin receptor. *Nature, 378*, 390–394. https://doi.org/10.1038/378390a0.
George, L., Chaverra, M., Todd, V., Lansford, R., & Lefcort, F. (2007). Nociceptive sensory neurons derive from contralaterally migrating, fate-restricted neural crest cells. *Nature Neuroscience, 10*, 1287–1293. https://doi.org/10.1038/nn1962.
George, L., Kasemeier-Kulesa, J., Nelson, B. R., Koyano-Nakagawa, N., & Lefcort, F. (2010). Patterned assembly and neurogenesis in the chick dorsal root ganglion. *The Journal of Comparative Neurology, 518*, 405–422. https://doi.org/10.1002/cne.22248.
Glenn Northcutt, R. (2005). The new head hypothesis revisited. *Journal of Experimental Zoology. Part B, Molecular and Developmental Evolution, 304*, 274–297. https://doi.org/10.1002/jez.b.21063.
Golding, J. P., Sobieszczuk, D., Dixon, M., Coles, E., Christiansen, J., Wilkinson, D., & Gassmann, M. (2004). Roles of erbB4, rhombomere-specific, and rhombomere-independent cues in maintaining neural crest-free zones in the embryonic head. *Developmental Biology, 266*, 361–372. https://doi.org/10.1016/j.ydbio.2003.11.003.
Graveson, A. C., Smith, M. M., & Hall, B. K. (1997). Neural crest potential for tooth development in a urodele amphibian: Developmental and evolutionary significance. *Developmental Biology, 188*, 34–42. https://doi.org/10.1006/dbio.1997.8563.
Green, S. A., Simoes-Costa, M., & Bronner, M. E. (2015). Evolution of vertebrates as viewed from the crest. *Nature, 520*, 474–482. https://doi.org/10.1038/nature14436.
Grenier, J., Teillet, M.-A., Grifone, R., Kelly, R. G., & Duprez, D. (2009). Relationship between neural crest cells and cranial mesoderm during head muscle development. *PLoS ONE, 4*, e4381. https://doi.org/10.1371/journal.pone.0004381.
Gross, J. B., & Hanken, J. (2008). Review of fate-mapping studies of osteogenic cranial neural crest in vertebrates. *Developmental Biology, 317*, 389–400. https://doi.org/10.1016/j.ydbio.2008.02.046.
Gross, J. B., & Hanken, J. (2005). Cranial neural crest contributes to the bony skull vault in adult Xenopus laevis: Insights from cell labeling studies. *Journal of Experimental zoology. Part B, Molecular and Developmental Evolution, 304*, 169–176. https://doi.org/10.1002/jez.b.21028.

Guidato, S., & Itasaki, N. (2007). Wise retained in the endoplasmic reticulum inhibits Wnt signaling by reducing cell surface LRP6. *Developmental Biology, 310*, 250–263. https://doi.org/10.1016/j.ydbio.2007.07.033.

Hall, B. K. (2009). Neuronal Cells and Nervous Systems. In B. K. Hall (Ed.). *The Neural Crest and Neural Crest Cells in Vertebrate Development and Evolution*Boston, MA: Springerhttps://doi.org/10.1007/978-0-387-09846-3_6.

Hall, B. K. (2008). The neural crest and neural crest cells: Discovery and significance for theories of embryonic organization. *Journal of Biosciences, 33*, 781–793. https://doi.org/10.1007/s12038-008-0098-4.

Hall, B. K. (2000). The neural crest as a fourth germ layer and vertebrates as quadroblastic not triploblastic. *Evolution & Development, 2*, 3–5. https://doi.org/10.1046/j.1525-142x.2000.00032.x.

Hall, B. K. (1998). Germ Layers and the Germ-Layer Theory Revisited. In M. K. Hecht, R. J. Macintyre, & M. T. Clegg (Vol. Eds.), *Evolutionary Biology. Evolutionary Biology. 30*(pp.)Boston, MA: Springer. https://doi.org/10.1007/978-1-4899-1751-5_5.

Halmi, C. A., Leonard, C. E., McIntosh, A. T., & Taneyhill, L. A. (2024). N-cadherin facilitates trigeminal sensory neuron outgrowth and target tissue innervation. *bioRxiv*. https://doi.org/10.1101/2024.05.20.594965.

Hamburger, V. (1961). Experimental analysis of the dual origin of the trigeminal ganglion in the chick embryo. *The Journal of Experimental Zoology, 148*, 91–123. https://doi.org/10.1002/jez.1401480202.

Harlow, D. E., & Barlow, L. A. (2007). Embryonic origin of gustatory cranial sensory neurons. *Developmental Biology, 310*, 317–328. https://doi.org/10.1016/j.ydbio.2007.07.042.

Harlow, D. E., Yang, H., Williams, T., & Barlow, L. A. (2011). Epibranchial placode-derived neurons produce BDNF required for early sensory neuron development. *Developmental Dynamics: an Official Publication of the American Association of Anatomists, 240*, 309–323. https://doi.org/10.1002/dvdy.22527.

Harris, M. L., & Erickson, C. A. (2007). Lineage specification in neural crest cell path-finding. *Developmental Dynamics: an Official Publication of the American Association of Anatomists, 236*, 1–19. https://doi.org/10.1002/dvdy.20919.

Hatta, K., Takagi, S., Fujisawa, H., & Takeichi, M. (1987). Spatial and temporal expression pattern of N-cadherin cell adhesion molecules correlated with morphogenetic processes of chicken embryos. *Developmental Biology, 120*, 215–227. https://doi.org/10.1016/0012-1606(87)90119-9.

Helms, J. A., & Schneider, R. A. (2003). Cranial skeletal biology. *Nature, 423*, 326–331. https://doi.org/10.1038/nature01656.

Hines, M. A., & Taneyhill, L. A. (2024). Elp1 function in placode-derived neurons is critical for proper trigeminal ganglion development. *Developmental Dynamics: an Official Publication of the American Association of Anatomists*. https://doi.org/10.1002/dvdy.749.

His, W. (1868). *Untersuchungen über die erste Anlage des Wirbelthierleibes: Die erste Entwickelung des Hühnchens im Ei / von Wilhelm His*. Leipzig: F.C.W. Vogel, https://doi.org/10.5962/bhl.title.15288.

Hochgreb-Hägele, T., & Bronner, M. E. (2013). A novel FoxD3 gene trap line reveals neural crest precursor movement and a role for FoxD3 in their specification. *Developmental Biology, 374*, 1–11. https://doi.org/10.1016/j.ydbio.2012.11.035.

Hong, C.-S., & Saint-Jeannet, J.-P. (2007). The activity of Pax3 and Zic1 regulates three distinct cell fates at the neural plate border. *Molecular Biology of the Cell, 18*, 2192–2202. https://doi.org/10.1091/mbc.e06-11-1047.

Hornbruch, A., Ma, G., Ballermann, M. A., Tumova, K., Liu, D., & Cairine Logan, C. (2005). A BMP-mediated transcriptional cascade involving Cash1 and Tlx-3 specifies first-order relay sensory neurons in the developing hindbrain. *Mechanisms of Development, 122*, 900–913. https://doi.org/10.1016/j.mod.2005.04.004.

Huang, C., Chan, J. A., & Schuurmans, C. (2014). Proneural bHLH genes in development and disease. *Current Topics in Developmental Biology, 110*, 75–127. https://doi.org/10.1016/B978-0-12-405943-6.00002-6.

Ito, K., Morita, T., & Sieber-Blum, M. (1993). In vitro clonal analysis of mouse neural crest development. *Developmental Biology, 157*, 517–525. https://doi.org/10.1006/dbio.1993.1154.

Ito, Y., Bringas, P., Jr, Mogharei, A., Zhao, J., Deng, C., & Chai, Y. (2002). Receptor-regulated and inhibitory Smads are critical in regulating transforming growth factor beta-mediated Meckel's cartilage development. *Developmental Dynamics: an Official Publication of the American Association of Anatomists, 224*, 69–78. https://doi.org/10.1002/dvdy.10088.

Jidigam, V. K., & Gunhaga, L. (2013). Development of cranial placodes: Insights from studies in chick. *Development, Growth & Differentiation, 55*, 79–95. https://doi.org/10.1111/dgd.12027.

Jin, E. J., Erickson, C. A., Takada, S., & Burrus, L. W. (2001). Wnt and BMP signaling govern lineage segregation of melanocytes in the avian embryo. *Developmental Biology, 233*, 22–37. https://doi.org/10.1006/dbio.2001.0222.

Johnston, M. C. (1966). A radioautographic study of the migration and fate of cranial neural crest cells in the chick embryo. *The Anatomical Record, 156*, 143–155. https://doi.org/10.1002/ar.1091560204.

Kasemeier-Kulesa, J. C., McLennan, R., Romine, M. H., Kulesa, P. M., & Lefcort, F. (2010). CXCR4 controls ventral migration of sympathetic precursor cells. *The Journal of Neuroscience, 30*, 13078–13088. https://doi.org/10.1523/JNEUROSCI.0892-10.2010.

Kenny, C., Dilshat, R., Seberg, H., Van Otterloo, E., Bonde, G., Helverson, A., ... Cornell, R. A. (2021). TFAP2 paralogs facilitate chromatin access for MITF at pigmentation genes but inhibit expression of cell-cell adhesion genes independently of MITF. *Plos Genetics*. https://doi.org/10.1101/2021.11.23.469757.

Kerosuo, L., & Bronner, M. E. (2016). CMyc regulates the size of the premigratory neural crest stem cell pool. *Cell Reports, 17*, 2648–2659. https://doi.org/10.1016/j.celrep.2016.11.025.

Kerosuo, L., & Bronner-Fraser, M. (2012). What is bad in cancer is good in the embryo: Importance of EMT in neural crest development. *Seminars in Cell & Developmental Biology, 23*, 320–332. https://doi.org/10.1016/j.semcdb.2012.03.010.

Knecht, A. K., & Bronner-Fraser, M. (2002). Induction of the neural crest: A multigene process. *Nature Reviews. Genetics, 3*, 453–461. https://doi.org/10.1038/nrg819.

Kondo, T., Sheets, P. L., Zopf, D. A., Aloor, H. L., Cummins, T. R., Chan, R. J., & Hashino, E. (2008). Tlx3 exerts context-dependent transcriptional regulation and promotes neuronal differentiation from embryonic stem cells. *Proceedings of the National Academy of Sciences of the United States of America, 105*, 5780–5785. https://doi.org/10.1073/pnas.0708704105.

Krispin, S., Nitzan, E., & Kalcheim, C. (2010). The dorsal neural tube: A dynamic setting for cell fate decisions. *Developmental Neurobiology, 70*, 796–812. https://doi.org/10.1002/dneu.20826.

Krull, C. E. (1998). Inhibitory interactions in the patterning of trunk neural crest migration. *Annals of the New York Academy of Sciences, 857*, 13–22. https://doi.org/10.1111/j.1749-6632.1998.tb10103.x.

Krull, C. E., Lansford, R., Gale, N. W., Collazo, A., Marcelle, C., Yancopoulos, G. D., ... Bronner-Fraser, M. (1997). Interactions of Eph-related receptors and ligands confer rostrocaudal pattern to trunk neural crest migration. *Current Biology: CB, 7*, 571–580. https://doi.org/10.1016/s0960-9822(06)00256-9.

Kulesa, P. M., Bailey, C. M., Kasemeier-Kulesa, J. C., & McLennan, R. (2010). Cranial neural crest migration: New rules for an old road. *Developmental Biology, 344*, 543–554. https://doi.org/10.1016/j.ydbio.2010.04.010.

Kulesa, P. M., & Fraser, S. E. (2000). In ovo time-lapse analysis of chick hindbrain neural crest cell migration shows cell interactions during migration to the branchial arches. *Development, 127*, 1161–1172. https://doi.org/10.1242/dev.127.6.1161.

Kulesa, P. M., & Fraser, S. E. (1998). Neural crest cell dynamics revealed by time-lapse video microscopy of whole embryo chick explant cultures. *Developmental Biology, 204*, 327–344. https://doi.org/10.1006/dbio.1998.9082.

Kulesa, P. M., & Gammill, L. S. (2010). Neural crest migration: Patterns, phases and signals. *Developmental Biology, 344*, 566–568. https://doi.org/10.1016/j.ydbio.2010.05.005.

Kuratani, S. C., Miyagawa-Tomita, S., & Kirby, M. L. (1991). Development of cranial nerves in the chick embryo with special reference to the alterations of cardiac branches after ablation of the cardiac neural crest. *Anatomy and Embryology, 183*, 501–514. https://doi.org/10.1007/bf00186439.

Kuriyama, S., & Mayor, R. (2008). Molecular analysis of neural crest migration. *Philosophical Transactions of the Royal Society of London. Series B, Biological Sciences, 363*, 1349–1362. https://doi.org/10.1098/rstb.2007.2252.

LaBonne, C., & Bronner-Fraser, M. (1998). Neural crest induction in Xenopus: Evidence for a two-signal model. *Development, 125*, 2403–2414. https://doi.org/10.1242/dev.125.13.2403.

Lassiter, R. N. T., Dude, C. M., Reynolds, S. B., Winters, N. I., Baker, C. V. H., & Stark, M. R. (2007). Canonical Wnt signaling is required for ophthalmic trigeminal placode cell fate determination and maintenance. *Developmental Biology, 308*, 392–406. https://doi.org/10.1016/j.ydbio.2007.05.032.

Le Douarin, N. (1982). *Developmental and cell biology series: The neural crest series number 12, developmental and cell biology series*. Cambridge, England: Cambridge University Press.

Le Douarin, N. (1973). A biological cell labeling technique and its use in expermental embryology. *Developmental Biology, 30*, 217–222. https://doi.org/10.1016/0012-1606(73)90061-4.

Le Douarin, N., Dulac, C., Dupin, E., & Cameron-Curry, P. (1991). Glial cell lineages in the neural crest. *Glia, 4*, 175–184. https://doi.org/10.1002/glia.440040209.

Le Douarin, N., & Kalcheim, C. (1999). *Developmental and cell biology series: The neural crest series number 36Developmental and cell biology series* (2nd ed.). Cambridge, England: Cambridge University Press.

Le Douarin, N. M. (1986). Cell line segregation during peripheral nervous system ontogeny. *Science, 231*, 1515–1522. https://doi.org/10.1126/science.3952494.

Le Douarin, N. M., & Dupin, E. (2014). *The neural crest, a fourth germ layer of the vertebrate embryo. Neural crest cells*. Elsevier 3–26. https://doi.org/10.1016/B978-0-12-401730-6.00001-6.

Le Douarin, N. M., & Dupin, E. (2018). The "beginnings" of the neural crest. *Developmental Biology, 444*, S3–S13. https://doi.org/10.1016/j.ydbio.2018.07.019.

Le Douarin, N. M., & Teillet, M. A. (1974). Experimental analysis of the migration and differentiation of neuroblasts of the autonomic nervous system and of neurectodermal mesenchymal derivatives, using a biological cell marking technique. *Developmental Biology, 41*, 162–184. https://doi.org/10.1016/0012-1606(74)90291-7.

Le Douarin, N. M., & Teillet, M. A. (1973). The migration of neural crest cells to the wall of the digestive tract in avian embryo. *Journal of Embryology and Experimental Morphology, 30*, 31–48.

Le Lièvre, C. S. (1978). Participation of neural crest-derived cells in the genesis of the skull in birds. *Journal of Embryology and Experimental Morphology, 47*, 17–37.

Lefcort, F., & George, L. (2007). Neural crest cell fate: To be or not to be prespecified. *Cell Adhesion Migration, 1*, 199–201. https://doi.org/10.4161/cam.1.4.5447.

Lefebvre, V., Li, P., & De Crombrugghe, B. (1998). A new long form of Sox5 (L-Sox5), Sox6 and Sox9 are coexpressed in chondrogenesis and cooperatively activate the type II collagen gene. *The EMBO Journal, 17*, 5718–5733. https://doi.org/10.1093/emboj/17.19.5718.

Lewis, J. L., Bonner, J., Modrell, M., Ragland, J. W., Moon, R. T., Dorsky, R. I., & Raible, D. W. (2004). Reiterated Wnt signaling during zebrafish neural crest development. *Development, 131*, 1299–1308. https://doi.org/10.1242/dev.01007.

Li, X., Chu, J., Wang, A., Zhu, Y., Chu, W. K., Yang, L., & Li, S. (2011). Uniaxial mechanical strain modulates the differentiation of neural crest stem cells into smooth muscle lineage on micropatterned surfaces. *PLoS ONE, 6*, e26029. https://doi.org/10.1371/journal.pone.0026029.

Lim, J., & Thiery, J. P. (2012). Epithelial-mesenchymal transitions: Insights from development. *Development, 139*, 3471–3486. https://doi.org/10.1242/dev.071209.

Ling, I. T. C., & Sauka-Spengler, T. (2019). Early chromatin shaping predetermines multipotent vagal neural crest into neural, neuronal and mesenchymal lineages. *Nature Cell Biology, 21*, 1504–1517. https://doi.org/10.1038/s41556-019-0428-9.

Logan, C., Wingate, R. J., McKay, I. J., & Lumsden, A. (1998). Tlx-1 and Tlx-3 homeobox gene expression in cranial sensory ganglia and hindbrain of the chick embryo: Markers of patterned connectivity. *The Journal of Neuroscience, 18*, 5389–5402.

Lukoseviciute, M., Gavriouchkina, D., Williams, R. M., Hochgreb-Hagele, T., Senanayake, U., Chong-Morrison, V., ... Sauka-Spengler, T. (2018). From pioneer to repressor: Bimodal foxd3 activity dynamically remodels neural crest regulatory landscape in vivo. *Developmental Cell, 47*, 608–628.e6. https://doi.org/10.1016/j.devcel.2018.11.009.

Lumsden, A., Clarke, J. D., Keynes, R., & Fraser, S. (1994). Early phenotypic choices by neuronal precursors, revealed by clonal analysis of the chick embryo hindbrain. *Development, 120*, 1581–1589. https://doi.org/10.1242/dev.120.6.1581.

Lumsden, A. G. (1988). Spatial organization of the epithelium and the role of neural crest cells in the initiation of the mammalian tooth germ. *Development, 103*, 155–169. https://doi.org/10.1242/dev.103.Supplement.155 (Suppl).

Lumsden, A. G., & Buchanan, J. A. (1986). An experimental study of timing and topography of early tooth development in the mouse embryo with an analysis of the role of innervation. *Archives of Oral Biology, 31*, 301–311. https://doi.org/10.1016/0003-9969(86)90044-0.

Lwigale, P. Y. (2001). Embryonic origin of avian corneal sensory nerves. *Developmental Biology, 239*, 323–337. https://doi.org/10.1006/dbio.2001.0450.

Ma, Q., Chen, Z., del Barco Barrantes, I., de la Pompa, J. L., & Anderson, D. J. (1998). Neurogenin1 is essential for the determination of neuronal precursors for proximal cranial sensory ganglia. *Neuron, 20*, 469–482. https://doi.org/10.1016/s0896-6273(00)80988-5.

Ma, Q., Kintner, C., & Anderson, D. J. (1996). Identification of neurogenin, a vertebrate neuronal determination gene. *Cell, 87*, 43–52. https://doi.org/10.1016/s0092-8674(00)81321-5.

Marchant, L., Linker, C., Ruiz, P., Guerrero, N., & Mayor, R. (1998). The inductive properties of mesoderm suggest that the neural crest cells are specified by a BMP gradient. *Developmental Biology, 198*, 319–329. https://doi.org/10.1016/s0012-1606(98)80008-0.

Martik, M. L., & Bronner, M. E. (2017). Regulatory logic underlying diversification of the neural crest. *Trends in Genetics: TIG, 33*, 715–727. https://doi.org/10.1016/j.tig.2017.07.015.

Martinsen, B. J., & Bronner-Fraser, M. (1998). Neural crest specification regulated by the helix-loop-helix repressor Id2. *Science, 281*, 988–991. https://doi.org/10.1126/science.281.5379.988.

Mayor, R., Guerrero, N., & Martínez, C. (1997). Role of FGF and noggin in neural crest induction. *Developmental Biology, 189*, 1–12. https://doi.org/10.1006/dbio.1997.8634.

Mayor, R., Morgan, R., & Sargent, M. G. (1995). Induction of the prospective neural crest of Xenopus. *Development, 121*, 767–777. https://doi.org/10.1242/dev.121.3.767.

Mayor, R., & Theveneau, E. (2014). The role of the non-canonical Wnt-planar cell polarity pathway in neural crest migration. *The Biochemical Journal, 457*, 19–26. https://doi.org/10.1042/BJ20131182.

Mayran, A., Khetchoumian, K., Hariri, F., Pastinen, T., Gauthier, Y., Balsalobre, A., & Drouin, J. (2018). Pioneer factor Pax7 deploys a stable enhancer repertoire for specification of cell fate. *Nature Genetics, 50*, 259–269. https://doi.org/10.1038/s41588-017-0035-2.

McKeown, S. J., Wallace, A. S., & Anderson, R. B. (2013). Expression and function of cell adhesion molecules during neural crest migration. *Developmental Biology, 373*, 244–257. https://doi.org/10.1016/j.ydbio.2012.10.028.

McLennan, R., Schumacher, L. J., Morrison, J. A., Teddy, J. M., Ridenour, D. A., Box, A. C., ... Kulesa, P. M. (2015). VEGF signals induce trailblazer cell identity that drives neural crest migration. *Developmental Biology, 407*, 12–25. https://doi.org/10.1016/j.ydbio.2015.08.011.

Mellott, D. O., & Burke, R. D. (2008). Divergent roles for Eph and ephrin in avian cranial neural crest. *BMC Developmental Biology, 8*, 56. https://doi.org/10.1186/1471-213X-8-56.

Milet, C., & Monsoro-Burq, A. H. (2012). Neural crest induction at the neural plate border in vertebrates. *Developmental Biology, 366*, 22–33. https://doi.org/10.1016/j.ydbio.2012.01.013.

Minoux, M., Holwerda, S., Vitobello, A., Kitazawa, T., Kohler, H., Stadler, M. B., & Rijli, F. M. (2017). Gene bivalency at polycomb domains regulates cranial neural crest positional identity. *Science, 355*, eaal2913. https://doi.org/10.1126/science.aal2913.

Minoux, M., & Rijli, F. M. (2010). Molecular mechanisms of cranial neural crest cell migration and patterning in craniofacial development. *Development, 137*, 2605–2621. https://doi.org/10.1242/dev.040048.

Moody, S. A., & Heaton, M. B. (1983c). Developmental relationships between trigeminal ganglia and trigeminal motoneurons in chick embryos. I. Ganglion development is necessary for motoneuron migration. *The Journal of Comparative Neurology, 213*, 327–343. https://doi.org/10.1002/cne.902130308.

Moody, S. A., & Heaton, M. B. (1983b). Developmental relationships between trigeminal ganglia and trigeminal motoneurons in chick embryos. II. Ganglion axon ingrowth guides motoneuron migration. *The Journal of Comparative Neurology, 213*, 344–349. https://doi.org/10.1002/cne.902130309.

Moody, S. A., & Heaton, M. B. (1983a). Developmental relationships between trigeminal ganglia and trigeminal motoneurons in chick embryos. III. Ganglion perikarya direct motor axon growth in the periphery. *The Journal of Comparative Neurology, 213*, 350–364. https://doi.org/10.1002/cne.902130310.

Moody, S. A., Quigg, M. S., & Frankfurter, A. (1989). Development of the peripheral trigeminal system in the chick revealed by an isotype-specific anti-beta-tubulin monoclonal antibody. *The Journal of Comparative Neurology*. https://doi.org/10.1002/cne.902790406.

Moore, R., & Larue, L. (2004). Cell surface molecules and truncal neural crest ontogeny: A perspective. *Birth Defects Research. Part C, Embryo Today: Reviews, 72*, 140–150. https://doi.org/10.1002/bdrc.20014.

Morikawa, Y., Zehir, A., Maska, E., Deng, C., Schneider, M. D., Mishina, Y., & Cserjesi, P. (2009). BMP signaling regulates sympathetic nervous system development through Smad4-dependent and -independent pathways. *Development, 136*, 3575–3584. https://doi.org/10.1242/dev.038133.

Nakagawa, S., & Takeichi, M. (1995). Neural crest cell-cell adhesion controlled by sequential and subpopulation-specific expression of novel cadherins. *Development, 121*, 1321–1332. https://doi.org/10.1242/dev.121.5.1321.

Nakamura, H., & Lievre, C. S. A.-L. (1982). Mesectodermal capabilities of the trunk neural crest of birds. *Development, 70*, 1–18. https://doi.org/10.1242/dev.70.1.1.

Newgreen, D. F. (1989). Physical influences on neural crest cell migration in avian embryos: Contact guidance and spatial restriction. *Developmental Biology, 131*, 136–148. https://doi.org/10.1016/s0012-1606(89)80045-4.

Newgreen, D. F., Gibbins, I. L., Sauter, J., Wallenfels, B., & Wütz, R. (1982). Ultrastructural and tissue-culture studies on the role of fibronectin, collagen and glycosaminoglycans in the migration of neural crest cells in the fowl embryo. *Cell and Tissue Research, 221*, 521–549. https://doi.org/10.1007/bf00215700.

Newgreen, D. F., & Gooday, D. (1985). Control of the onset of migration of neural crest cells inavian embryos. Role of Ca++-dependent cell adhesions. *Cell and Tissue Research, 239*, 329–336. https://doi.org/10.1007/bf00218011.

Newgreen, D. F., Scheel, M., & Kastner, V. (1986). Anatomical and experimental studies on early morphogenesis of the avian perichordal region: Differential effect of notochordal chondroitinase-sensitive material on neural crest and sclerotome cells. *Cell and Tissue Research.*

Noden, D. M. (1975). An analysis of migratory behavior of avian cephalic neural crest cells. *Developmental Biology, 42*, 106–130. https://doi.org/10.1016/0012-1606(75)90318-8.

Olesnicky Killian, E. C., Birkholz, D. A., & Artinger, K. B. (2009). A role for chemokine signaling in neural crest cell migration and craniofacial development. *Developmental Biology, 333*, 161–172. https://doi.org/10.1016/j.ydbio.2009.06.031.

Olsson, L., & Hanken, J. (1996). Cranial neural-crest migration and chondrogenic fate in the oriental fire-bellied toad Bombina orientalis: Defining the ancestral pattern of head development in anuran amphibians. *Journal of Morphology, 229*, 105–120. https://doi.org/10.1002/(SICI)1097-4687(199607)229:13.0.CO;2-2.

Osborne, N. J., Begbie, J., Chilton, J. K., Schmidt, H., & Eickholt, B. J. (2005). Semaphorin/neuropilin signaling influences the positioning of migratory neural crest cells within the hindbrain region of the chick. *Developmental Dynamics: an Official Publication of the American Association of Anatomists, 232*, 939–949. https://doi.org/10.1002/dvdy.20258.

Perez-Alcala, S., Nieto, M. A., & Barbas, J. A. (2004). LSox5 regulates RhoB expression in the neural tube and promotes generation of the neural crest. *Development, 131*, 4455–4465. https://doi.org/10.1242/dev.01329.

Pieper, M., Eagleson, G. W., Wosniok, W., & Schlosser, G. (2011). Origin and segregation of cranial placodes in Xenopus laevis. *Developmental Biology, 360*, 257–275. https://doi.org/10.1016/j.ydbio.2011.09.024.

Piloto, S., & Schilling, T. F. (2010). Ovo1 links Wnt signaling with N-cadherin localization during neural crest migration. *Development, 137*, 2603. https://doi.org/10.1242/dev.055913.

Pla, P., Moore, R., Morali, O. G., Grille, S., Martinozzi, S., Delmas, V., & Larue, L. (2001). Cadherins in neural crest cell development and transformation. *Journal of Cellular Physiology, 189*, 121–132. https://doi.org/10.1002/jcp.10008.

Potterf, S. B., Mollaaghababa, R., Hou, L., Southard-Smith, E. M., Hornyak, T. J., Arnheiter, H., & Pavan, W. J. (2001). Analysis of SOX10 function in neural crest-derived melanocyte development: SOX10-dependent transcriptional control of dopachrome tautomerase. *Developmental Biology, 237*, 245–257. https://doi.org/10.1006/dbio.2001.0372.

Prasad, M. S., Sauka-Spengler, T., & LaBonne, C. (2012). Induction of the neural crest state: Control of stem cell attributes by gene regulatory, post-transcriptional and epigenetic interactions. *Developmental Biology, 366*, 10–21. https://doi.org/10.1016/j.ydbio.2012.03.014.

Revenu, C., Streichan, S., Donà, E., Lecaudey, V., Hufnagel, L., & Gilmour, D. (2014). Quantitative cell polarity imaging defines leader-to-follower transitions during collective migration and the key role of microtubule-dependent adherens junction formation. *Development, 141*, 1282–1291. https://doi.org/10.1242/dev.101675.

Richardson, J., Gauert, A., Briones Montecinos, L., Fanlo, L., Alhashem, Z. M., Assar, R., ... Linker, C. (2016). Leader cells define directionality of trunk, but not cranial, neural crest cell migration. *Cell Reports, 15*, 2076–2088. https://doi.org/10.1016/j.celrep.2016.04.067.

Rickmann, M., Fawcett, J. W., & Keynes, R. J. (1985). The migration of neural crest cells and the growth of motor axons through the rostral half of the chick somite. *Development, 90*, 437–455. https://doi.org/10.1242/dev.90.1.437.

Rinon, A., Molchadsky, A., Nathan, E., Yovel, G., Rotter, V., Sarig, R., & Tzahor, E. (2011). P53 coordinates cranial neural crest cell growth and epithelial-mesenchymal transition/delamination processes. *Development, 138*, 1827–1838. https://doi.org/10.1242/dev.053645.

Roellig, D., Tan-Cabugao, J., Esaian, S., & Bronner, M. E. (2017). Dynamic transcriptional signature and cell fate analysis reveals plasticity of individual neural plate border cells. *Elife, 6*. https://doi.org/10.7554/elife.21620.

Rogers, C. D., Saxena, A., & Bronner, M. E. (2013). Sip1 mediates an E-cadherin-to-N-cadherin switch during cranial neural crest EMT. *The Journal of Cell Biology, 203*, 835–847. https://doi.org/10.1083/jcb.201305050.

Rothstein, M., Bhattacharya, D., & Simoes-Costa, M. (2018). The molecular basis of neural crest axial identity. *Developmental Biology, 444*(1), S170–S180. https://doi.org/10.1016/j.ydbio.2018.07.026.

Saint-Jeannet, J.-P., & Moody, S. A. (2014). Establishing the pre-placodal region and breaking it into placodes with distinct identities. *Developmental Biology, 389*, 13–27. https://doi.org/10.1016/j.ydbio.2014.02.011.

Sasselli, V., Pachnis, V., & Burns, A. J. (2012). The enteric nervous system. *Developmental Biology, 366*, 64–73. https://doi.org/10.1016/j.ydbio.2012.01.012.

Sauka-Spengler, T., & Bronner-Fraser, M. (2008). A gene regulatory network orchestrates neural crest formation. *Nature Reviews. Molecular Cell Biology, 9*, 557–568. https://doi.org/10.1038/nrm2428.

Scarpa, E., Szabó, A., Bibonne, A., Theveneau, E., Parsons, M., & Mayor, R. (2015). Cadherin switch during EMT in neural crest cells leads to contact inhibition of locomotion via repolarization of forces. *Developmental Cell, 34*, 421–434. https://doi.org/10.1016/j.devcel.2015.06.012.

Scerbo, P., & Monsoro-Burq, A. H. (2020). The vertebrate-specific VENTX/NANOG gene empowers neural crest with ectomesenchyme potential. *Science Advances, 6*. https://doi.org/10.1126/sciadv.aaz1469 eaaz1469.

Schilling, T. F., & Kimmel, C. B. (1994). Segment and cell type lineage restrictions during pharyngeal arch development in the zebrafish embryo. *Development, 120*, 483–494. https://doi.org/10.1242/dev.120.3.483.

Schlosser, G., & Northcutt, R. G. (2000). Development of neurogenic placodes in Xenopus laevis. *The Journal of Comparative Neurology, 418*, 121–146. https://doi.org/10.1002/(sici)1096-9861(20000306)418:23.3.co;2-d.

Schoenwolf, G. C., Bleyl, S. B., Brauer, P. R., & Francis-West, P. H. (2009). *Development of the peripheral nervous system. Larsen's human embryology*. Elsevier 297–318. https://doi.org/10.1016/b978-0-443-06811-9.10010-7.

Schroeder, T. E. (1970). Neurulation in Xenopus laevis. An analysis and model based upon light and electron microscopy. *Journal of Embryology and Experimental Morphology, 23*, 427–462.

Schwarz, Q., Vieira, J. M., Howard, B., Eickholt, B. J., & Ruhrberg, C. (2008). Neuropilin 1 and 2 control cranial gangliogenesis and axon guidance through neural crest cells. *Development, 135*, 1605–1613. https://doi.org/10.1242/dev.015412.

Seberg, H. E., Van Otterloo, E., Loftus, S. K., Liu, H., Bonde, G., Sompallae, R., ... Cornell, R. A. (2017). TFAP2 paralogs regulate melanocyte differentiation in parallel with MITF. *PLoS Genetics, 13*, e1006636. https://doi.org/10.1371/journal.pgen.1006636.

Sela-Donenfeld, D., & Kalcheim, C. (1999). Regulation of the onset of neural crest migration by coordinated activity of BMP4 and Noggin in the dorsal neural tube. *Development, 126*, 4749–4762. https://doi.org/10.1242/dev.126.21.4749.

Serbedzija, G. N., Bronner-Fraser, M., & Fraser, S. E. (1992). Vital dye analysis of cranial neural crest cell migration in the mouse embryo. *Development, 116*, 297–307. https://doi.org/10.1242/dev.116.2.297.

Shellard, A., & Mayor, R. (2021). Collective durotaxis along a self-generated stiffness gradient in vivo. *Nature, 600*, 690–694. https://doi.org/10.1038/s41586-021-04210-x.

Shellard, A., & Mayor, R. (2019). Integrating chemical and mechanical signals in neural crest cell migration. *Current Opinion in Genetics & Development, 57*, 16–24. https://doi.org/10.1016/j.gde.2019.06.004.

Shiau, C. E., & Bronner-Fraser, M. (2009). N-cadherin acts in concert with Slit1-Robo2 signaling in regulating aggregation of placode-derived cranial sensory neurons. *Development, 136*, 4155–4164. https://doi.org/10.1242/dev.034355.

Shiau, C. E., Lwigale, P. Y., Das, R. M., Wilson, S. A., & Bronner-Fraser, M. (2008). Robo2-Slit1 dependent cell-cell interactions mediate assembly of the trigeminal ganglion. *Nature Neuroscience, 11*, 269–276. https://doi.org/10.1038/nn2051.

Shigetani, Y., & Itasaki, N. (2007). Expression of Wise in chick embryos. *Developmental Dynamics: an Official Publication of the American Association of Anatomists, 236*, 2277–2284. https://doi.org/10.1002/1097-0177(20010301)220:33.3.co;2-x.

Shyamala, K., Yanduri, S., Girish, H. C., & Murgod, S. (2015). Neural crest: The fourth germ layer. *Journal of Oral and Maxillofacial Pathology: JOMFP. 19*, 221–229. https://doi.org/10.4103/0973-029X.164536.

Sieber-Blum, M. (1989). Commitment of neural crest cells to the sensory neuron lineage. *Science, 243*, 1608–1611. https://doi.org/10.1126/science.2564699.

Sieber-Blum, M., & Cohen, A. M. (1980). Clonal analysis of quail neural crest cells: They are pluripotent and differentiate in vitro in the absence of noncrest cells. *Developmental Biology, 80*, 96–106. https://doi.org/10.1016/0012-1606(80)90501-1.

Simoes-Costa, M., & Bronner, M. E. (2016). Reprogramming of avian neural crest axial identity and cell fate. *Science, 352*, 1570–1573. https://doi.org/10.1126/science.aaf2729.

Simões-Costa, M., & Bronner, M. E. (2015). Establishing neural crest identity: A gene regulatory recipe. *Development, 142*, 242–257. https://doi.org/10.1242/dev.105445.

Smith, A., & Graham, A. (2001). Restricting Bmp-4 mediated apoptosis in hindbrain neural crest. *Developmental Dynamics: an Official Publication of the American Association of Anatomists, 220*, 276–283. https://doi.org/10.1002/1097-0177(20010301)220:3 < 276::aid-dvdy1110 > 3.3.co;2-x.

Smith, A., Robinson, V., Patel, K., & Wilkinson, D. G. (1997). The EphA4 and EphB1 receptor tyrosine kinases and ephrin-B2 ligand regulate targeted migration of branchial neural crest cells. *Current Biology: CB, 7*, 561–570. https://doi.org/10.1016/s0960-9822(06)00255-7.

Smith, M. M., & Hall, B. K. (1990). Development and evolutionary origins of vertebrate skeletogenic and odontogenic tissues. *Biological Reviews of the Cambridge Philosophical Society, 65*, 277–373. https://doi.org/10.1111/j.1469-185x.1990.tb01427.x.

Soldatov, R., Kaucka, M., Kastriti, M. E., Petersen, J., Chontorotzea, T., Englmaier, L., ... Adameyko, I. (2019). Spatiotemporal structure of cell fate decisions in murine neural crest. *Science, 364*, eaas9536. https://doi.org/10.1126/science.aas9536.

Sommer, L., Ma, Q., & Anderson, D. J. (1996). neurogenins, a novel family of atonal-related bHLH transcription factors, are putative mammalian neuronal determination genes that reveal progenitor cell heterogeneity in the developing CNS and PNS. *Molecular and Cellular Neurosciences, 8*, 221–241. https://doi.org/10.1006/mcne.1996.0060.

Square, T. A., Jandzik, D., Massey, J. L., Romášek, M., Stein, H. P., Hansen, A. W., ... Medeiros, D. M. (2020). Evolution of the endothelin pathway drove neural crest cell diversification. *Nature, 585*, 563–568. https://doi.org/10.1038/s41586-020-2720-z.

Stark, M. R., Sechrist, J., Bronner-Fraser, M., & Marcelle, C. (1997). Neural tube-ectoderm interactions are required for trigeminal placode formation. *Development, 124*, 4287–4295. https://doi.org/10.1242/dev.124.21.4287.

Steventon, B., Araya, C., Linker, C., Kuriyama, S., & Mayor, R. (2009). Differential requirements of BMP and Wnt signalling during gastrulation and neurulation define two steps in neural crest induction. *Development, 136*, 771–779. https://doi.org/10.1242/dev.029017.

Steventon, B., Carmona-Fontaine, C., & Mayor, R. (2005). Genetic network during neural crest induction: From cell specification to cell survival. *Seminars in Cell & Developmental Biology, 16*, 647–654. https://doi.org/10.1016/j.semcdb.2005.06.001.

Steventon, B., Mayor, R., & Streit, A. (2014). Neural crest and placode interaction during the development of the cranial sensory system. *Developmental Biology, 389*, 28–38. https://doi.org/10.1016/j.ydbio.2014.01.021.

Streit, A. (2002). Extensive cell movements accompany formation of the otic placode. *Developmental Biology, 249*, 237–254. https://doi.org/10.1006/dbio.2002.0739.

Stuhlmiller, T. J., & García-Castro, M. I. (2012). Current perspectives of the signaling pathways directing neural crest induction. *Cellular and Molecular Life Sciences: CMLS, 69*, 3715–3737. https://doi.org/10.1007/s00018-012-0991-8.

Szabó, A., Theveneau, E., Turan, M., & Mayor, R. (2019). Neural crest streaming as an emergent property of tissue interactions during morphogenesis. *PLoS Computational Biology, 15*, e1007002. https://doi.org/10.1371/journal.pcbi.1007002.

Takahashi, M., & Osumi, N. (2005). Identification of a novel type II classical cadherin: Rat cadherin19 is expressed in the cranial ganglia and Schwann cell precursors during development. *Developmental Dynamics: an Official Publication of the American Association of Anatomists, 232*, 200–208. https://doi.org/10.1002/dvdy.20209.

Takano-Maruyama, M., Chen, Y., & Gaufo, G. O. (2012). Differential contribution of Neurog1 and Neurog2 on the formation of cranial ganglia along the anterior-posterior axis. *Developmental Dynamics: an Official Publication of the American Association of Anatomists, 241*, 229–241. https://doi.org/10.1002/dvdy.22785.

Takeda, K., Yasumoto, K., Takada, R., Takada, S., Watanabe, K., Udono, T., ... Shibahara, S. (2000). Induction of melanocyte-specific microphthalmia-associated transcription factor by Wnt-3a. *The Journal of Biological Chemistry, 275*, 14013–14016. https://doi.org/10.1074/jbc.c000113200.

Taneyhill, L. A., Coles, E. G., & Bronner-Fraser, M. (2007). Snail2 directly represses cadherin6B during epithelial-to-mesenchymal transitions of the neural crest. *Development, 134*, 1481–1490. https://doi.org/10.1242/dev.02834.

Teraoka, M. E., Paschaki, M., Muta, Y., & Ladher, R. K. (2009). Rostral paraxial mesoderm regulates refinement of the eye field through the bone morphogenetic protein (BMP) pathway. *Developmental Biology, 330*, 389–398. https://doi.org/10.1016/j.ydbio.2009.04.008.

Thawani, A., & Groves, A. K. (2020). Building the border: Development of the chordate neural plate border region and its derivatives. *Frontiers in Physiology, 11*, 608880. https://doi.org/10.3389/fphys.2020.608880.

Théveneau, E., Duband, J.-L., & Altabef, M. (2007). Ets-1 confers cranial features on neural crest delamination. *PLoS ONE, 2*, e1142. https://doi.org/10.1371/journal.pone.0001142.

Theveneau, E., Marchant, L., Kuriyama, S., Gull, M., Moepps, B., Parsons, M., & Mayor, R. (2010). Collective chemotaxis requires contact-dependent cell polarity. *Developmental Cell, 19*, 39–53. https://doi.org/10.1016/j.devcel.2010.06.012.

Theveneau, E., & Mayor, R. (2012). Neural crest delamination and migration: From epithelium-to-mesenchyme transition to collective cell migration. *Developmental Biology, 366*, 34–54. https://doi.org/10.1016/j.ydbio.2011.12.041.

Theveneau, E., Steventon, B., Scarpa, E., Garcia, S., Trepat, X., Streit, A., & Mayor, R. (2013). Chase-and-run between adjacent cell populations promotes directional collective migration. *Nature Cell Biology, 15*, 763–772. https://doi.org/10.1038/ncb2772.

Thiery, J. P., Duband, J. L., & Delouvée, A. (1982). Pathways and mechanisms of avian trunk neural crest cell migration and localization. *Developmental Biology, 93*, 324–343. https://doi.org/10.1016/0012-1606(82)90121-x.

Thiery, J. P., & Sleeman, J. P. (2006). Complex networks orchestrate epithelial-mesenchymal transitions. *Nature Reviews. Molecular Cell Biology, 7*, 131–142. https://doi.org/10.1038/nrm1835.

Thomas, A. J., & Erickson, C. A. (2009). FOXD3 regulates the lineage switch between neural crest-derived glial cells and pigment cells by repressing MITF through a non-canonical mechanism. *Development, 136*, 1849–1858. https://doi.org/10.1242/dev.031989.

Tosney, K. W. (2004). Long-distance cue from emerging dermis stimulates neural crest melanoblast migration. *Developmental Dynamics: an Official Publication of the American Association of Anatomists, 229*, 99–108. https://doi.org/10.1002/dvdy.10492.

Tosney, K. W. (1982). The segregation and early migration of cranial neural crest cells in the avian embryo. *Developmental Biology, 89*, 13–24. https://doi.org/10.1016/0012-1606(82)90289-5.

Trainor, P. A., Melton, K. R., & Manzanares, M. (2003). Origins and plasticity of neural crest cells and their roles in jaw and craniofacial evolution. *The International Journal of Developmental Biology, 47*, 541–553.

Trainor, P. A., & Tam, P. P. (1995). Cranial paraxial mesoderm and neural crest cells of the mouse embryo: Co-distribution in the craniofacial mesenchyme but distinct segregation in branchial arches. *Development, 121*, 2569–2582. https://doi.org/10.1242/dev.121.8.2569.

Urrutia, H. A., Stundl, J., & Bronner, M. E. (2024). Tlx3 mediates neuronal differentiation and proper condensation of the developing trigeminal ganglion. *Developmental Biology, 515*, 79–91. https://doi.org/10.1016/j.ydbio.2024.07.005.

Vallin, J., Thuret, R., Giacomello, E., Faraldo, M. M., Thiery, J. P., & Broders, F. (2001). Cloning and characterization of three Xenopus slug promoters reveal direct regulation by Lef/beta-catenin signaling. *The Journal of Biological Chemistry, 276*, 30350–30358. https://doi.org/10.1074/jbc.M103167200.

Ventéo, S., Desiderio, S., Cabochette, P., Deslys, A., Carroll, P., & Pattyn, A. (2019). Neurog2 deficiency uncovers a critical period of cell fate plasticity and vulnerability among neural-crest-derived somatosensory progenitors. *Cell Reports, 29*, 2953–2960.e2. https://doi.org/10.1016/j.celrep.2019.11.002.

Wang, H. U., & Anderson, D. J. (1997). Eph family transmembrane ligands can mediate repulsive guidance of trunk neural crest migration and motor axon outgrowth. *Neuron, 18*, 383–396. https://doi.org/10.1016/s0896-6273(00)81240-4.

Wang, W.-D., Melville, D. B., Montero-Balaguer, M., Hatzopoulos, A. K., & Knapik, E. W. (2011). Tfap2a and Foxd3 regulate early steps in the development of the neural crest progenitor population. *Developmental Biology, 360*, 173–185. https://doi.org/10.1016/j.ydbio.2011.09.019.

Weston, J. A. (1963). A radioautographic analysis of the migration and localization of trunk neural crest cells in the chick. *Developmental Biology, 6*, 279–310. https://doi.org/10.1016/0012-1606(63)90016-2.

Weston, J. A., & Butler, S. L. (1966). Temporal factors affecting localization of neural crest cells in the chicken embryo. *Developmental Biology, 14*, 246–266. https://doi.org/10.1016/0012-1606(66)90015-7.

Williams, R. M., Candido-Ferreira, I., Repapi, E., Gavriouchkina, D., Senanayake, U., Ling, I. T. C., ... Sauka-Spengler, T. (2019). Reconstruction of the global neural crest gene regulatory network in vivo. *Developmental Cell, 51*, 255–276.e7. https://doi.org/10.1016/j.devcel.2019.10.003.

Williams, R. M., Lukoseviciute, M., Sauka-Spengler, T., & Bronner, M. E. (2022). Single-cell atlas of early chick development reveals gradual segregation of neural crest lineage from the neural plate border during neurulation. *Elife, 11*. https://doi.org/10.7554/eLife.74464.

Wu, C.-Y., Hooper, R. M., Han, K., & Taneyhill, L. A. (2014). Migratory neural crest cell αN-catenin impacts chick trigeminal ganglia formation. *Developmental Biology, 392*, 295–307. https://doi.org/10.1016/j.ydbio.2014.05.016.

Wu, C.-Y., & Taneyhill, L. A. (2019). Cadherin-7 mediates proper neural crest cell-placodal neuron interactions during trigeminal ganglion assembly. *Genesis, 57*, e23264. https://doi.org/10.1002/dvg.23264.

Xu, H., Dude, C. M., & Baker, C. V. H. (2008). Fine-grained fate maps for the ophthalmic and maxillomandibular trigeminal placodes in the chick embryo. *Developmental Biology, 317*, 174–186. https://doi.org/10.1016/j.ydbio.2008.02.012.

Xu, Y., Lopes, C., Qian, Y., Liu, Y., Cheng, L., Goulding, M., ... Ma, Q. (2008). Tlx1 and Tlx3 coordinate specification of dorsal horn pain-modulatory peptidergic neurons. *The Journal of Neuroscience, 28*, 4037–4046. https://doi.org/10.1523/JNEUROSCI.4126-07.2008.

Yntema, C. L. (1944). Experiments on the origin of the sensory ganglia of the facial nerve in the chick. *The Journal of Comparative Neurology, 81*, 147–167. https://doi.org/10.1002/cne.900810204.

Yook, J. I., Li, X.-Y., Ota, I., Hu, C., Kim, H. S., Kim, N. H., ... Weiss, S. J. (2006). A Wnt-Axin2-GSK3beta cascade regulates Snail1 activity in breast cancer cells. *Nature Cell Biology, 8*, 1398–1406. https://doi.org/10.1038/ncb1508.

York, J. R., & McCauley, D. W. (2020). The origin and evolution of vertebrate neural crest cells. *Open Biology, 10*, 190285. https://doi.org/10.1098/rsob.190285.

Young, N. M., Hu, D., Lainoff, A. J., Smith, F. J., Diaz, R., Tucker, A. S., ... Marcucio, R. S. (2014). Embryonic bauplans and the developmental origins of facial diversity and constraint. *Development, 141*, 1059–1063. https://doi.org/10.1242/dev.099994.

Yu, H.-H., & Moens, C. B. (2005). Semaphorin signaling guides cranial neural crest cell migration in zebrafish. *Developmental Biology, 280*, 373–385. https://doi.org/10.1016/j.ydbio.2005.01.029.

Zalc, A., Sinha, R., Gulati, G. S., Wesche, D. J., Daszczuk, P., Swigut, T., ... Wysocka, J. (2021). Reactivation of the pluripotency program precedes formation of the cranial neural crest. *Science, 371*. https://doi.org/10.1126/science.abb4776.

Zhu, Y., Li, X., Janairo, R. R. R., Kwong, G., Tsou, A. D., Chu, J. S., ... Li, S. (2019). Matrix stiffness modulates the differentiation of neural crest stem cells in vivo. *Journal of Cellular Physiology, 234*, 7569–7578. https://doi.org/10.1002/jcp.27518.

CHAPTER FOUR

Collective migration in complex environments: The case of the trunk neural crest

Hanna-Maria Häkkinen[1], Soraya Villaseca[1], Marie De La Burgade[1], Zainalabdeen Alhashem[1], and Elena Scarpa*

Department of Physiology, Development and Neuroscience, University of Cambridge, Cambridge, United Kingdom
*Corresponding author. e-mail address: es697@cam.ac.uk

Contents

1. Introduction 110
2. Brief history of trunk neural crest research 112
3. The trunk and cranial neural crest migratory pathways 113
   3.1 Cranial neural crest migratory pathway 113
4. Trunk neural crest migratory pathways 114
   4.1 The ventromedial pathway 115
   4.2 The dorsolateral pathway 115
5. Cell fates of the trunk neural crest 116
   5.1 Models of fate acquisition in the TNC 117
   5.2 Mechanisms of fate acquisition in the zebrafish ventromedial TNC streams 118
   5.3 A role for TNC spatial context for cell fate specification? 120
6. Modes of neural crest migration 121
   6.1 Collective cell migration in NCCs 121
   6.2 Collective and individual cell migration in TNCs 124
7. Patterning of trunk NC migration and interaction with surrounding environment 127
   7.1 Trunk neural crest interactions with somites 128
   7.2 Interactions with the extracellular matrix 131
   7.3 Physical cues and NC migration 134
8. In vitro models of neural crest: advantages and limitations 135
9. Conclusion 138
References 139

Abstract

The neural crest is a highly invasive, multipotent embryonic cell population common to all vertebrates. Neural crest cells migrate all along the anteroposterior axis of the vertebrate embryos, crossing complex microenvironments during their journey and eventually halting their migration to give rise to a variety of derivatives. At cranial levels,

[1] All authors contributed equally to this work.

Current Topics in Developmental Biology, Volume 164
ISSN 0070-2153, https://doi.org/10.1016/bs.ctdb.2025.01.004

neural crest cells originate cartilage and bone of the skull and face, cranial ganglia and glia and pigment cells. In contrast, neural crest of the trunk is unable to form ecto-mesenchymal tissues such as cartilage and bone, but instead contributes to the cardiac outflow tract, enteric neurons, sensory and sympathetic neurons, Schwann cells and pigment across the vertebrate trunk. Defects in neural crest formation and migration can result in an array of birth defects and childhood malignancies collectively known as neurocristopathies, and investigation of the mechanisms underlying neural crest migration has significant clinical relevance. Considerable progress has been made in recent years in our understanding of the principles underlying collective cell migration of cranial neural crest cells. However, the extracellular environment trunk neural crest traverse *in vivo* is radically different from that experienced by cranial neural crest cells. Here, we review collective cell migration, fate specification and current *in vivo* and *in vitro* models of trunk neural crest migration under the lens of the complex interaction of this extraordinary cell population with its complex tissue environment.

1. Introduction

The neural crest is a transient, multipotent cell population specific to all vertebrates. Neural crest cells give rise to an extraordinarily diverse array of cell types, from cartilage and bone, smooth muscle and cardiac valves, pigment cells, cranial ganglia, endocrine cells and neurons and glia of the peripheral nervous system. Because of their versatility, they are sometimes described as the fourth germ layer, and it has been hypothesized that the evolution of the neural crest has significantly contributed to the success and diversification across the vertebrate clade (Gans & Northcutt, 1983).

Neural crest cells are important for successful embryonic development, and human mutations affecting neural crest formation, migration and differentiation are among the most common birth defects in live-born infants. Collectively, diseases associated with neural crest developmental defects are named neurocristopathies (Etchevers, Amiel, & Lyonnet, 2007; Trainor, 2016). Neural crest cells are induced and specified at the dorsal edge of the neural tube and from there they migrate ventrally towards their target destinations. The gene regulatory networks underlying neural crest induction and specification have been studied in detail (for reviews: Betancur, Bronner-Fraser, & Sauka-Spengler, 2010; Martik & Bronner, 2017; Sauka-Spengler & Bronner-Fraser, 2008).

Neural crest cells are endowed with extraordinary migratory and invasive capabilities and cross a variety of environments during their developmental journey. Indeed, neural crest cells are present along the whole anteroposterior axis of vertebrate embryos, from head to tail. We can distinguish cranial, vagal, trunk and sacral subpopulations of neural crest, based on their axial level of origin in the developing embryo (Le Douarin, 1999). Depending on their anteroposterior location, these cells traverse distinct tissue environments as they migrate, and differ in their fate, mode of migration and how they interact with their surroundings.

Indeed, cranial neural crest (CNCs) populations migrate through the head mesenchyme as large collectives of cells, and they are unique in their ability to differentiate into ectomesenchyme derivatives such as cartilage and bone of the craniofacial skeleton. In addition, they form cranial ganglia and associated glial cells, as well as pigment cells (Knight & Schilling, 2006; Szabo & Mayor, 2018). This cell population is characterized by collective migration in streams to specific regions of the head and neck of the embryo. They migrate through well-defined pathways, often between segments of surrounding tissue patterned by repulsive cues (Gammill, Gonzalez, & Bronner-Fraser, 2007; Osborne, Begbie, Chilton, Schmidt, & Eickholt, 2005). These repulsive cues appear not to be prepatterned, but instead they dynamically emerge as a result of collective interactions between neural crest cells and surrounding placodal tissues (Szabo et al., 2018).

Vagal, trunk and sacral neural crest subpopulations migrate in streams of varying shape and size through strikingly denser, complex tissue environments and do not originate ectomesenchyme. The Vagal Neural Crest originates from the dorsal aspect of the neural tube at the level of the future neck (between somites 1 to 7 in the chick embryo) and forms the cardiac outflow tract, enteric neurons, smooth muscle and ganglia of the lung and pancreas (Hutchins et al., 2018). The Sacral Neural Crest, in contrast, is originated from the neural tube at caudal levels (posterior to somite 28 in the chick) and contributes to enteric neurons of the hindgut (Serbedzija, Burgan, Fraser, & Bronner-Fraser, 1991; Shepherd & Eisen, 2011; Wang, Chan, Sham, Burns, & Chan, 2011) and to glial cells of the bladder (Wiese, Deal, Ireland, Cantrell, & Southard-Smith, 2017).

The cellular principles underlying collective neural crest migration have been extensively studied using chick and frog and are very well understood in these model organisms (Szabo & Mayor, 2018). Significant advances have been made in recent years in our understanding of the cell biology underlying the collective migration of the neural crest. Indeed, we now understand fairly well: (i) cell-cell interactions and cell biological principles of collective movement of CNCs; (ii) guidance (long range and short range); (iii) the dynamic interaction of CNCs with surrounding tissues and recent advances have highlighted an important role for physical properties of the environment for CNC collective migration.

On the other hand, whether the same cell biological mechanisms underlie the collective migration of the trunk neural crest (TNCs) is currently less understood, possibly because of organism-specific variations in the organization of the trunk tissue environment and the patterning guidance cues surrounding the migrating trunk neural crest. In this book chapter, we will summarize the current knowledge around *in vivo* and *in vitro* modes of TNC development and migration, cell fate specification, and interaction with the surrounding tissue environment of the trunk neural crest and highlight exciting future directions and open questions in the field.

2. Brief history of trunk neural crest research

In 1868, His first described the neural crest as a "cord" between neural and non neural ectoderm and called it "ganglionic crest" recognizing that this tissue later gave rise to trunk ganglia. Julia Platt, in 1893, carried out observations in amphibian embryos and was one of the first investigators to recognize that cranial neural crest originates head ectomesenchyme. Later microsurgery experiments in chick and amphibia, which involved the physical removal of sections of neural tube along the embryonic trunk (Hammond & Yntema, 1947; Yntema & Hammond, 1945; Yntema & Hammond, 1954;) established that trunk neural crest cells were instead able to give rise to sympathetic, chromaffin and dorsal root ganglia [for a historical review of the early years of neural crest research, see (Le Douarin & Dupin, 2018)].

Early studies were mainly focused on understanding which tissues the neural crest originates from, as given the transient and migratory nature of the neural crest, it was virtually impossible to follow their migration at the time. Lineage tracing following injection of H^3 -thymidine in dividing neuroepithelial cells by Weston (Noden, 1975; Weston & Butler, 1966) in chick and amphibia by Chibon (Chibon, 1964; Chibon, 1967) allowed grafting of labeled chunk of tissues in unlabeled recipients. However, tritiated thymidine gets diluted by proliferation, making long-term cell tracking difficult. The development of the chick-quail chimaera method by Nicole LeDouarin (reviewed in Le Douarin, 1999) revolutionized neural crest lineage tracing, allowing the unequivocally identify neural crest migratory routes and fates. By xenografting quail neural crest tissue into chick host and taking advantage of an easily recognizable heterochromatic region in the nucleoli of quail, but not chick, tissues, it became possible to identify single migrating neural crest and neural crest derivatives in chick host embryos.

The quail-chick chimaera system, together with Dii labelinglabeling of single cells pioneered by Marianne Bronner and Scott Fraser, allowed live imaging and lineage tracing of neural crest cells in avian and fish embryos (Fraser & Bronner-Fraser, 1991; Raible & Eisen, 1994) was instrumental in broadening our understanding of cranial and trunk neural crest migratory pathways.

3. The trunk and cranial neural crest migratory pathways

3.1 Cranial neural crest migratory pathway

In avian embryos, cranial neural crest at prosencephalic, mesencephalic and rhombencephalic leves migrate subectodermally, in a relatively cell-free space, and only a few cells enter the underlying head mesoderm (Fig. 1A) (Le Lievre & Le Douarin, 1974). At rhombencephalic levels, the cranial neural crest coalesces in defined streams (reviewed in Szabo & Mayor, 2018).

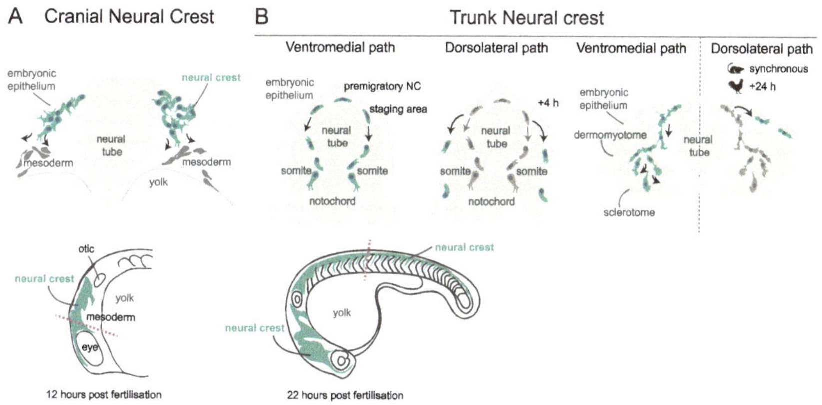

Fig. 1 Cranial neural crest and trunk neural crest migratory pathways. (A) Cross section view through the cranial region of a 12 hpf zebrafish embryo, just posterior to the eye as depicted in the bottom panel. (B) Trunk neural crest migratory pathways. A cross-section through the zebrafish trunk is represented on the left hand side panel, whilst the right hand side panel represents a cross section through the trunk of a 30 ss chick embryo.

Very similar to avian models, in rat and mouse embryos (Serbedzija, Bronner-Fraser, & Fraser, 1992; Tan & Morriss-Kay, 1986), the neural crest emigrate subectodermally from the neural plate border when the neural tube is still open. *Xenopus* CNCs also form lateral to the neural plate and migrate subectodermally over the relatively loose underlying head mesoderm whilst the anterior neural plate is open (Barriga, Franze, Charras, & Mayor, 2018; Sadaghiani & Thiebaud, 1987). Similarly, zebrafish CNC migrates ventrally from their site of origin lateral to the anterior neural keel (Fig. 1A) (Schilling & Kimmel, 1994). Collective co-migration of *Xenopus* and zebrafish CNC and neighboring epibranchial placodes is important for patterning the placodes into separate tissue subunits (Theveneau et al., 2013). In zebrafish embryos, CNC migrates in a nearly completely cell-free space over a thin string (1–2 cells wide) of paraxial head mesoderm (Schilling & Kimmel, 1994) (Fig. 1).

Despite species-specific differences, all cranial neural crests activate an epithelial-to-mesenchymal transition program that endows the cells with migratory and invasive capabilities (reviewed in Gouignard, Andrieu, & Theveneau, 2018; Theveneau & Mayor, 2012).

Cranial neural crest migrates as large collectives of cells, interacting with one another via several cell-cell adhesion molecules and progressively dispersing as they migrate ventrally. In addition, in zebrafish, chick (Richardson et al., 2016) and *Xenopus* (Kuriyama et al., 2014) cranial neural crest appear to be able to freely exchange neighbors, and they do not show permanent leader or follower cell identities.

4. Trunk neural crest migratory pathways

Trunk neural crest migrates from the dorsal neural tube through two main migratory paths, the ventromedial path and a dorsolateral path (Fig. 1B and C). TNCs migration is initiated in a rostrocaudal order so that anterior cells migrate at earlier developmental stages than posterior ones. In contrast with CNC, trunk neural crest is exposed to a strikingly different tissue environment throughout their developmental migration. Indeed, the trunk neural migratory paths are strictly confined through dense tissues, and the physical environment they traverse has a complex extracellular matrix composition. This aspect was already highlighted in 1989 by Newgreen, who investigated the physical restriction of TNC cells during their migration across the chick embryo trunk (Newgreen, 1989).

4.1 The ventromedial pathway

In the trunk, neural crest can take two main migratory paths, the ventromedial path and the dorsolateral path (Fig. 1B and C). Before entering either of the paths, premigratory trunk cells sit in a so-called "staging area" dense in extracellular matrix at the dorsal aspect of the neural tube (Wehrle-Haller & Weston, 1997). From the migration staging area, cells entering the ventromedial path first squeeze in the interstitial space between the neural tube and somite (Fig. 1B and C). This pathway is highly conserved across species, though differences in the morphogenesis of the somites between amniotes and anamniotes embryos may account for some of the differences in the modalities of TNC migration across model organisms. Indeed, whilst the chick and mouse somite tissue is composed of dermomyotome and a loosely organized sclerotome (Gossler & Hrabe De Angelis, 1998), in zebrafish, *Xenopus* and other anamniotes the sclerotome represents only a very small fraction of the somite. Indeed, in anamniotes the somite is mostly composed of a densely packed muscle precursor tissue called myotome (Keller, 2000; Stickney, Barresi, & Devoto, 2000). In mouse and chick embryos, TNC starts migrating through the ventral path between the early somites and neural tube in a non-segmented manner. Once the somite has maturated into a sclerotome and dermomyotome, cells intermingle between the anterior region of the sclerotome and the basal membrane of the dermomyotome (Le Douarin, 1999) (Fig. 1C). In chick embryos, the ventromedial path was first observed by quail-chick chimaera methods (Le Douarin & Teillet, 1974), and vital DiI labeling later confirmed similar routes of migration between chick and mouse (Serbedzija, Bronner-Fraser, & Fraser, 1989; Serbedzija, Fraser, & Bronner-Fraser, 1990). Whilst in avians and mice models where the TNC migrates on the anterior half of the somite (Fig. 3), in amphibians, the NC migrates on the posterior region of the somite, whilst in zebrafish, only one stream is established per each somite (Raible et al., 1992; Richardson et al., 2016) and the stream aligns in the middle of the somite (Raible et al., 1992). Across vertebrates, TNC of the ventromedial path is recognized to originate sympathetic and sensory neurons, chromaffin cells, glia of the peripheral nervous system and pigment.

4.2 The dorsolateral pathway

A subset of trunk neural crest cells migrates utilizing the dorsolateral pathway, entering a restricted space between the ectoderm and somites (Fig. 1B and C). The timing of dorsolateral migration differs among vertebrates: in mice, dorsolateral migration occurs simultaneously with the ventromedial invasion (Serbedzija et al., 1990), while in chick, it occurs 24 h after ventromedial migration starts (Serbedzija et al., 1989). Similar to the chick model, in zebrafish, entry and migration of neural crest cells from the premigratory "staging area" into the dorsolateral pathway occurs approximately four hours later than entry and migration in the ventromedial pathway (Raible et al., 1992). In *Xenopus,* there are notable differences in the trunk neural crest migratory pathways. Indeed, most cells either take a ventral pathway circumnavigating the neural tube and notochord or a dorsal pathway into the dorsal fin and dorsal pigment stripe. In contrast with other model organisms, only a small fraction of presumptive pigment cells takes the dorsolateral pathway (Collazo, Bronner-Fraser, & Fraser, 1993; Krotoski & Bronner-Fraser, 1986; Sadaghiani & Thiebaud, 1987). Across species, neural crests migrating along the dorsolateral pathway are fated to become pigment cells.

5. Cell fates of the trunk neural crest

The trunk neural crest is an embryonic stem cell population which gives rise to a wide array of cell types at specific embryonic locations. Many of these cell types are essential to vertebrates. Much remains to be known about the mechanisms underlying how these diverse fates are produced. Understanding TNC fate specification would have direct consequences on our understanding of vertebrate development and TNC-derived diseases such as neuroblastoma and Hirschsprung's disease (Etchevers et al., 2007; Vega-Lopez, Cerrizuela, & Aybar, 2017).

In contrast to cranial neural crest, which can give rise to ectomesenchymal derivatives such as cartilage and bone of the head (reviewed in detail in and in Szabo & Mayor, 2018), trunk neural crest cannot form ectomesenchyme derivatives. This was first demonstrated thanks to the development of the chick-quail chimaera method (Le Douarin, 1999). By carrying out heterotopic and heterochronic transplantation experiments Le Douarin and colleagues observed that, whilst quail trunk neural crest transplanted into the rhombencephalon were able to co-migrate with endogenous cranial neural crest (Le Douarin, Teillet, & Le Lievre, 1977; Nakamura & Ayer-Le Lievre, 1982) thus retaining the capacity to respond to cues within the head environment, they were unable to contribute to facial cartilage and skeleton. In a complementary study, the grafting of cranial neural crest tissue into trunk dorsal neural tube resulted instead into ectopic ectomesenchyme tissue at the trunk level (Noden, 1983).

5.1 Models of fate acquisition in the TNC

TNC fate specification is intrinsically tied to specific embryonic locations after migration. A first wave of cells migrates ventromedially and gives rise to sympathetic neurons, sensory neurons of the dorsal root ganglia, pigment cells and glia, whilst a second wave of cells later starts a dorsolateral migration and becomes pigment cells. Though species specific differences are present across the vertebrate clade a combination of intrinsic regulatory programs and environmental signals are necessary for NCC differentiation (reviewed in Rocha, Singh, Ahsan, Beiriger, & Prince, 2020).

It remains uncertain how the different fates of the TNC are acquired, and whether the fates are determined before, during, or after migration. Several models have been proposed to explain fate specification in the TNC for a review see (Dawes & Kelsh, 2021):

The direct fate restriction model postulated TNCs to be fully multipotent throughout migration and require environmental cues to get specified at their final location (Bronner-Fraser & Fraser, 1988, 1989; Collazo, Bronner-Fraser, & Fraser, 1993). This was suggested by early findings in chick and frog, as lineage tracing studies using Dii injections or transplantation of fluorescently labeled TNCCs revealed that the majority of labeled TNCs originated heterogeneous clones with multiple derivative cell types.

Another model, called the progressive fate restriction model eventually became the more classical way to interpret TNC specification. It was conceived as a series of restrictive choices that diminish TNCC potential before and during migration and influence their differentiation toward specific lineages (Anderson, 1989; Klymkowsky, Cortez Rossi, & Artinger, 2010; Simoes-Costa & Bronner, 2016) . Early on, distinct neurogenic and melanogenic sublineages of TNCs were proposed (Henion & Weston, 1997). In mouse TNCs, bipotent TNCCs appeared biassed towards a sensory neuroglial fate, specified by the transcription factor Neurog2, followed by a second decision separating autonomic (phox2bb/prrx1 positive cells) from mesenchymal lineages (Zirlinger, Lo, Mcmahon, Mcmahon, & Anderson, 2002). In this model, cells go through an initial coactivation, before gradually biassing towards a particular fate, and eventually commit to a certain program during migration. Supporting this model, a recent study using single cell RNA-seq of mouse NC cells detected fate biased NC already during delamination from the neural tube (Soldatov et al., 2019). TNCCs are transcriptionally diverse, and it was recently shown in zebrafish that there is already expression of genes associated with xanthophore pigment lineages at the pre-migratory stage (Lencer, Prekeris, & Artinger, 2021). However, expression of fate markers at transcriptional level does not necessarily results in fate restriction, and functional assays are necessary to unequivocally establish the potency of a stem cell.

A third model, cyclical fate restriction, has been recently proposed to combine seemingly contradictory insights into TNC fate specification. The cyclical fate restriction model posits that cells remain highly multipotent while also co-expressing different fate factors throughout migration but are only restricted dynamically through environmental choices (Kelsh et al., 2021). It was demonstrated using single cell RNA-seq as well as multiplexed fluorescent in situ hybridization that zebrafish TNCs co-express key fate specification genes such as the iridophore marker Ltk as well as the early sympathetic neuron fate marker phox2bb without partial fate restriction (Subkhankulova et al., 2023). This suggests that zebrafish TNC largely retain pluripotency throughout migration, and that specific fates choices are made directly, but dynamically, from a highly multipotent state. In support of the concept that TNC retain multipotency whilst migrating, an elegant study using Cre-mediated recombination of the Confetti multicolor lineage tracing system in mice (Snippert & Clevers, 2011) revealed TNC cells to be multipotent *in vivo* at both premigratory and migratory stages (Baggiolini et al., 2015). However, because of the difficulty of tracking cell fates *in vivo* in live mouse embryo, it remains uncertain how and when the different fates are acquired (Baggiolini et al., 2015).

5.2 Mechanisms of fate acquisition in the zebrafish ventromedial TNC streams

Zebrafish TNC migrate in a similar way to other vertebrates (Raible et al., 1992). As in many other organisms, TNC derivatives are sympathetic neurons, sensory neurons (DRG), glial cells, and different subtypes of pigment cells (Raible & Eisen, 1994).

Early lineage tracing studies of fluorescently labeled premigratory TNC cells (Raible & Eisen, 1994) suggested that whilst some of the earlier ventromedially migrating TNC were multipotent and able to originate multiple cell types (sympathetic neurons, DRG and glia), most of TNC appeared to be lineage restricted, generating progenitors that gave rise to one or two cell types. Similarly to all other vertebrates, lineage tracing of dorsolaterally migrating cells revealed that they give rise to pigment derivatives, though three different pigment cell types are present in zebrafish (melanocytes, iridophores, and xanthophores). Heterochronic transplantation experiments of early migrating and late migrating TNC cells (Raible & Eisen, 1996) revealed plasticity in the ventromedial TNC subpopulation, with late migrating cells acquiring the ability to originate DRG when early migrating cells were ablated. More recently, experiments where the early migrating cells were ablated using a laser also supported regulative interactions between premigratory cells, although the long term fates of cells were not investigated (Richardson et al., 2016).

Live imaging studies using the tissue-specific promoter Sox10, revealed that for TNCCs leading migration along the ventral path, Notch signaling might play a central role in determining sympathetic fates in zebrafish TNC, by deciding migratory identity and thus determining early and late migrating cells. TNC medial migration in zebrafish is directed by a single larger and faster leader cell, which divides halfway through migration (Richardson et al., 2016). The leaders undergo an asymmetric cell division that separates migratory identity between the distal daughter becoming a new leader, and the proximal daughter behaving as a follower (Alhashem, Camargo-Sosa, Kelsh, & Linker, 2022). The distal daughter becomes the new leader, and eventually clonally forms the sympathetic ganglia, while the proximal daughter gives rise to Schwann cells. A possible explanation underlying this division is the asymmetric distribution of Notch signaling controls identity allocation during the leader's division, maintaining a single leader in the chain (Alhashem, Feldner-Busztin et al., 2022).

Notch regulates the cell cycle of the TNC, which is necessary for TNC migration. Cells with higher Notch activity quickly progress through the G1/S transition and remain in S-phase longer. Leaders are larger and faster than followers (Richardson et al., 2016), and keep higher Notch activity as they migrate. In line with this observation, the strength of Notch signaling is influenced by the area of cell-cell contact, with smaller cells being more likely to act as signal-sending cells. Notch activity, in turn, regulates the expression of the sympathetic fate specification factor *phox2bb* (Ernsberger et al., 1995; Pei et al., 2013) . Indeed, the expression of Phox2bb is higher in leaders than in follower cells (Alhashem, Feldner-Busztin et al., 2022).

In line with observations in zebrafish, Notch has been suggested to be a pivotal factor in neuroglial fate determination in the mouse and avian DRG(Cornell & Eisen, 2005; Wiszniak & Schwarz, 2019) . In the DRG, cells expressing Delta-like 1 (DLL1) interact with their neighboring cells through lateral inhibition. DLL1-expressing cells adopt a neuronal fate, while signaling other cells to inhibit their neuronal differentiation and instead go through glial differentiation, increasing the expression of glial markers sox10 and PFAB7(Morrison et al., 2000; Wakamatsu, Maynard, & Weston, 2000). Cells expressing Notch1 are at the border of the ganglia, while cells expressing neuronal markers (Hu, Delta) are in the center (Wakamatsu et al., 2000). In zebrafish, by 36 h post-fertilization (hpf), the most lateral DRG cells begin to express neurogenin 1, resulting in a decrease in Sox10 expression, while other DRG cells continue to differentiate as glia, as evidenced by the expression of Foxd3 (Nitzan, Pfaltzgraff, Labosky, & Kalcheim, 2013). In quail, a role for glucose metabolism and nutrient input has also been recently reported to influence differentiation programs. Glucose intake promoted neural and melanoblast differentiation, while pyruvate intake promoted myofibroblasts (Nekooie Marnany et al., 2023).

5.3 A role for TNC spatial context for cell fate specification?

The microenvironment in which NCCs migrate significantly influences their fate determination. As previously discussed, different models exist to attempt to explain the role of spatial context in TNC fate specification: in the DFR model, fates would be driven by signals in post migratory environments, while in the CFR model, NCCs become biased towards specific fates by local environmental signals (Subkhankulova et al., 2023).

TNCC are also known to influence the fates of other tissues. As they migrate under the dermomyotome of the somites, they express Notch and other signals influencing the mesoderm to future fates and promote muscle differentiation (Rios, Serralbo, Salgado, & Marcelle, 2011). In addition, zebrafish TNC are required for the maintenance of the embryonic hematopoietic stem cell niche, and loss of TNC results in loss of HSCs (Damm & Clements, 2017). During the gene co-activation phases, TNCC might be able to respond to environmental inputs, so, probably, the somites and other tissues the TNC migrates next to impact TNC differentiation as well (Erickson, Kameneva, & Adameyko, 2023; Soldatov et al., 2019). Indeed, in zebrafish Neuregulin 1 and 2 expressed by spinal cord motorneurons and somite cells adjacent to migrating TNC are required for ERBB3 signaling and formation of DRG and sympathetic neurons as they provide a signal for migrating TNC cells to pause their migration at the DRG level (Honjo, Kniss, & Eisen, 2008).

External mechanical signals play a crucial role in tissue shaping by influencing cell morphology, directing cell migration, and altering intercellular relationships (Chacon-Martinez, Koester, & Wickstrom, 2018). Within this context, leader cells exhibit heightened polarization and large protrusions, which are sustained by their distal daughter cells as they become the new leaders (Alhashem, Camargo-Sosa et al., 2022; Alhashem, Feldner-Busztin et al., 2022; Richardson et al., 2016). The shape of these leaders may be affected by their confined surroundings, specifically in the interstitial space between the notochord, neural tube, and somites. This dynamic environment leads to distinct migratory paths for trunk neural crest cells (TNCCs): medial early migrating TNCCs and lateral late migrating TNCCs follow different trajectories, resulting in varied fates depending on the extent of their migration (Alhashem, Camargo-Sosa et al., 2022; Raible & Eisen, 1994; Raible & Eisen, 1996). Thus, the relationship between TNCC migratory identity and terminal differentiation reveals a coherent pattern in trunk neural crest migration and fate allocation. Whether the physical properties of the surrounding tissue environment impact on dynamic fate choices of TNC has not yet been explored.

6. Modes of neural crest migration

Depending on their anteroposterior positioning within the vertebrate embryo, neural crest cells migrate as large, coordinated collectives, as streams or as individual cells. Modes of migration across neural crest subpopulations are influenced by both intrinsic (from the NCCs themselves) and extrinsic factors (from the environment).

6.1 Collective cell migration in NCCs

Cranial neural crest cells migrate as large cell collectives. In collective cell migration NCCs move as a cohesive group while maintaining transient cell-cell contacts. Collective migration allows for coordinated movement of a large number of cells and can be faster and more efficient in guiding cells to their final position (Mayor & Etienne-Manneville, 2016). Cell-cell interactions mechanically link cells to one another and mediate the transmission of polarizing signals across the cell population, thus cells will influence the movement of each other. During collective cell migration, cadherins and other cell adhesion molecules play a key role in maintaining the polarity of neural crest cell collectives, a process known as "Contact Inhibition of Locomotion" (Carmona-Fontaine et al., 2008; Kotini et al., 2018; Scarpa et al., 2015).

Contact inhibition of locomotion is a cell-cell interaction behavior first observed by Abercrombie and Heaysman in 1953 in chick heart fibroblast and defined as the ability of a cell to change the direction of its motion after contact with another cell (Abercrombie & Heaysman, 1953). In frog CNC, CIL manifests as a contact-dependent cell polarity, so that the activity of the small GTPase RhoA is found at the cell-cell contact (Carmona-Fontaine et al., 2008) whilst the active form of the small GTPase Rac1, which promotes polymerization of branched actin networks, lamellipodia formation and integrin-ECM adhesion (Ridley et al., 2003) is polarized at the free edge of NC cells (Theveneau et al., 2010), thus promoting cell migration away from the site of cell-cell contact and CNC dispersion away from one another (Carmona-Fontaine et al., 2008; Scarpa et al., 2015).

After NC cells have been induced and specified (reviewed in Sauka-Spengler & Bronner-Fraser, 2008) they delaminate from the neural tube and undergo an epithelial-to-mesenchymal transition (EMT) by initiating expression of Snail1/2 and Twist as pro-EMT transcription factors in concert with neural crest specifiers FoxD3, Sip1Zeb2 and Sox9/10. The activation of the EMT gene regulatory network results in NCCs switching their cadherins expression. As a result of the EMT program, all NCCs undergo a process called "cadherin switch", where high levels of type I cadherins (E-Cadherin, N-Cadherin) are downregulated and type II cadherins become expressed instead (Nieto, 2009; Rogers, Saxena, & Bronner, 2013).

CNCs express E-Cadherin at premigratory stages in avians (Dady, Blavet, & Duband, 2012; Rogers et al., 2013), mice (Breau, Pietri, Stemmler, Thiery, & Weston, 2008), frog and fish (Barriga, Maxwell, Reyes, & Mayor, 2013; Barriga et al., 2018; Scarpa et al., 2015) and switch to expressing low levels of N-Cadherin during migratory stages.

On the other hand, TNC never express E-Cadherin. Instead, they switch from expressing weak levels of N-Cadherin compared to neural tube to upregulating a cascade of Type II cadherins. Indeed, an initial expression of cadherin6B in TNC tissue is thought to aid its delamination from neural tissue (Akitaya & Bronner-Fraser, 1992; Duband, Volberg, Sabanay, Thiery, & Geiger, 1988; Nakagawa & Takeichi, 1995). Cadherin-6b is subsequently downregulated via Snail2 transcriptional regulation (Coles, Taneyhill, & Bronner-Fraser, 2007; Taneyhill, Coles, & Bronner-Fraser, 2007) and Cadherin7 and 11 are activated (Chalpe, Prasad, Henke,

& Paulson, 2010; Cheung et al., 2005). This process appears to be conserved across vertebrates. Further details of trunk NC delamination cascade have been extensively reviewed in (Gouignard et al., 2018; Theveneau & Mayor, 2012).

The downregulation of high levels of type I Cadherins increases the fluidity of the tissue so that NCCs transform from epithelial non-motile cells to mesenchymal cells with motile capabilities (Kuriyama et al., 2014). The "cadherin switch" is essential to acquire the Contact Inhibition of Locomotion (CIL) behavior, as loss of E-Cadherin and expression of N-Cadherin and other mesenchymal cadherins like Cadherin6B (Clay & Halloran, 2014) promote cell separation and EMT via contact-dependent cell polarity (Carmona-Fontaine et al., 2008; Scarpa et al., 2015). In addition, the maintenance of a pro-migratory phenotype is aided by PDGF-A/PDGFRα signaling by controlling N-cadherin upregulation after EMT (Bahm, et al., 2017). In *Xenopus*, the onset of CNC EMT is promoted by changes in tissue mechanics of the underlying head mesoderm. Indeed, convergent-extension movements of the mesoderm cause an increase in tissue stiffness, which stimulates EMT, the E- to N-Cadherin switch and consequent tissue softening of CNC (Barriga et al., 2018; Marchant, Malmi-Kakkada, Espina, & Barriga, 2022) thus promoting the onset of neural crest migration.

In summary, after delamination, the CNC invade the space between the epidermal and mesodermal layers as a collective sheet. Because CIL acts as a driving force promoting neural crest dispersion, additional interactions within the CNC itself, and with chemotactic cues secreted by surrounding tissues are essential for collective tissue guidance.

Despite EMT and the onset of CIL promoting cell dispersion, CNC are nevertheless able to maintain group migration thanks to a co-attraction loop, whereby they respond to a chemoattractant, the C3 factor of the complement cascade, in an autocrine manner (Carmona-Fontaine et al., 2011). Both, C3 and C3aR, the G protein-coupled receptor for C3a, are expressed in migrating CNC cells. C3a/C3aR signaling leads to increased Rac1 activity at the leading edge of CNCs, promoting protrusion stability and attraction of CNCs towards one another. This mechanism ensures that cells within CNC collectives continue to loosely interact with one another despite CIL.

How are CNCs guided to their destinations? Attractive guidance cues, such as Sdf-1 (also known as CXCL12) secreted by tissues surrounding the CNC (Theveneau et al., 2010) can attract neural crest cells and reinforce the directionality of the moving cell collective. Indeed, CNC are guided towards tissue sources of Sdf-1 thanks to a "collective chemotaxis" mechanism. Indeed, CNC cells express the Sdf-1 receptor Cxcr4. Activation of the Cxcr4 signaling pathway results in local activation of Rac1 at the leading edge of cells, which causes stabilization of leading cell protrusions and directional cell migration (Theveneau et al., 2010). It is important to remark that the CNC chemotactic response to Sdf-1 is contact dependent: indeed, whilst isolated CNC are not able to respond to Sdf-1, in NCCs clusters reinforce contact-dependent polarization of Rac1 at the leading edge of lamellipodial protrusions, enhancing the directionality of collective cell migration.

CIL, co-attraction and collective chemotaxis have been initially described by Mayor and colleagues in the frog embryo and are the bases for our understanding of CNC collective cell migration in frogs. Although species-specific differences may exist, it is not always possible to draw comparisons, as conservation of mechanisms across species has not been explored systematically. For example, in chick embryos, VEGF has been described as a key chemoattractant for CNC migration (Mclennan, Teddy, Kasemeier-Kulesa, Romine, & Kulesa, 2010), however, the requirement for Sdf-1 has not been explored in chick embryos, and whilst PDGF, which is closely related to VEGF, has been found to be required for chemotaxis in frog (Bahm, et al., 2017), a requirement for VEGF in frogs has never been explored.

6.2 Collective and individual cell migration in TNCs

During collective migration, cells located at the leading edge of the group often act as leaders directing the movement of the group, while the rest act as followers. Leader cells are often characterized by larger, highly polarized lamellipodial protrusion, higher speed and directionality of migration (Mayor & Etienne-Manneville, 2016). *In vivo*, leader cells may lead migration through complex and challenging environments and require a metabolic profile that allows them to support energy demanding tasks (Emtenani et al., 2022; Garde & Sherwood, 2021). CNC collectives in frog, chick and zebrafish are characterized by high tissue fluidity, with cells often exchanging neighbors and leaders and often exchanging positions during migration (Kuriyama et al., 2014; Richardson et al., 2016). In zebrafish embryos, laser ablation experiments have demonstrated that leader cells are not a fixed population, as ablation of leaders has no detrimental effect on collective cell migration; instead, follower cells take the leading position and guide collective cell migration (Richardson et al., 2016).

On the other hand, in zebrafish trunk neural crest cells of the ventromedial pathway migrate as a single chain of cells, with one chain per each somite (Fig. 2). The cells appear to be loosely connected via transient cell-cell interactions, similar to CNCs. Indeed, the type II cadherin Pchd10a is required to maintain cell-cell interactions and integrity of the chain of migrating cells in zebrafish (Williams, Hsu, Rossi, & Artinger, 2018). A larger, faster and more directional leader cell directs the migration. Zebrafish TNC migration is very stereotyped and requires leader cells to direct the movement (Richardson et al., 2016). Both identities, leaders and followers, appear to be acquired before the initiation of TNC migration and contribute to intrinsic differences between cranial and trunk in the zebrafish (Alhashem, Feldner-Busztin et al., 2022; Richardson et al., 2016). Indeed, laser ablation of leader cells stops completely the ventral migration of TNC cells, evidencing that followers cannot exchange positions with leader cells and direct the migration of the group (Richardson et al., 2016). However, ablation of a single follower cell results in leader cells stalling forward movement until contact with another follower is re-established,

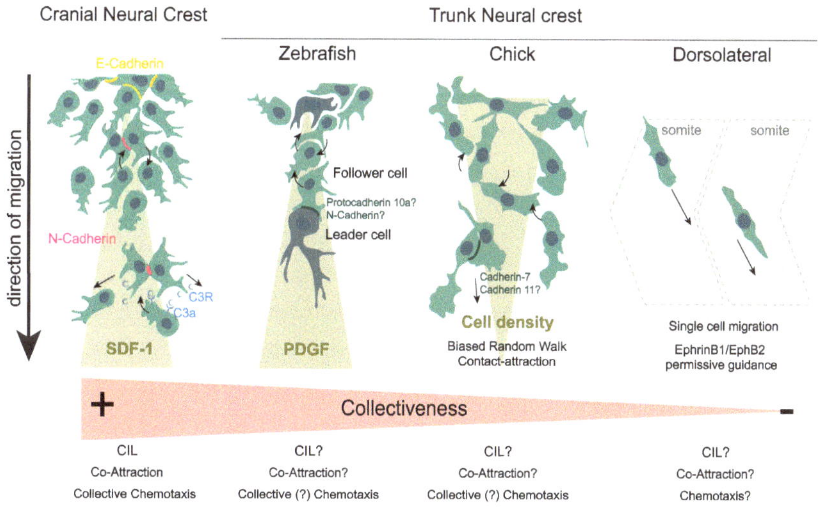

Fig. 2 Modes of migration across different neural crest cell populations. From left to right: overview of CNC migration. Stream collective migration of zebrafish trunk neural crest cells. Biased random walk of chick trunk neural crest cells. Single cell confined migration of dorsolateral neural crest cells, represented as observed in the zebrafish.

demonstrating that whilst only leader cells direct the migration of the group, cell-cell interactions are essential to confer directionality to the stream of migrating cells in a manner reminiscent of CNC contact-dependent cell polarity. These findings suggest that TNC cells experience Contact Inhibition of Locomotion. Indeed, live imaging of both zebrafish and chick TNC cells in their native environment reveals that TNCs appear to cease protrusive activity at the point of cell-cell contact (Li et al., 2019; Richardson et al., 2016). When chick TNC cells lose cell-cell contact, cells repolarize towards the direction of migration and the movement is only restored when cells establish contact between each other again, suggesting that cell-cell contact is crucial for TNC cell migration (Li et al., 2019). Importantly, most of the surface proteins known to mediate CIL in CNC cells, such as Par3, N-cadherin and Cadherin11 (Becker, Mayor, & Kashef, 2013; Chalpe et al., 2010), are also expressed in TNC cells (Moore et al., 2013), although their role in mediating CIL in trunk neural crest cells has not yet been directly tested. However, the putative Shh receptor *cdon*, a direct binding partner of N-Cadherin (Kang, Feinleib, Knox, Ketteringham, & Krauss, 2003; Lu & Krauss, 2010), is cell-autonomously required for zebrafish TNC migration and correct polarization of lamellipodial protrusions by maintaining N-Cadherin plasma membrane localization, thus suggesting that the cellular mechanisms underlying contact-dependent cell polarity are conserved across CNC and TNCs (Powell et al., 2015).

In contrast to collective cell migration, some NCCs appear to migrate as individual cells (Fig. 2). A cell migrating individually with high cell-ECM adhesion on a 2D substrate is characterized by a front-to-back cell polarity with a lamellipodium protruding at the leading edge and a retractile tail at its back. This polarity is controlled by the distribution of the active form of small GTPases Rho A, which accumulates at the trailing edge and of Rac1/Cdc42, prominently located at the leading edge of a migrating cell. These in turn, impact cytoskeleton organization and interactions with the ECM (Ridley et al., 2003). A molecular clutch mechanism where protrusions are driven by actin filament assembly at the leading edge and engagement of cell matrix adhesions promotes exertion of a traction force movement on the ECM, allowing forward motion of cells (Elosegui-Artola, Trepat, & Roca-Cusachs, 2018). In contrast, when single cells migrate in confined environments or under low cell-ECM adhesion conditions, they acquire an ameboid mode of migration, where locomotion is guided by high cortical contractility, membrane blebbing and non-specific interactions with the surrounding microenvironment (reviewed in Paluch, Aspalter, & Sixt, 2016).

A recent study exploited cultured thick tissue sections of the chick trunk to live image fluorescently labeled TNCs in conditions as close as possible to their *in vivo* environment (Li et al., 2019). The authors focused on the ventromedially migrating chick TNCs that invade the loose tissue of the sclerotome. Quantitative analysis of cell motion over time revealed that these cells undergo a biased random motion, migrating most of their time as individuals, with highly polarized lamellipodia directed towards the ventral axis of the embryo (Fig. 2). The directionality of chick TNC migration appears to be regulated by cell density, as cells undergo a "contact-attraction" behavior where they form transient cell-cell interactions, often co-migrating together for some time before repolarizing their lamellipodia in different directions in a manner reminiscent of contact inhibition of locomotion. This results in lower directionality and rounder cell shapes in dorsal, higher density regions where cells interact more often with one another and in more persistent migration in ventral lower density regions. Possibly, chemotactic guidance mediated by Sdf-1, secreted by the dorsal aorta may also contribute to the higher directionality of some of these loose TNC collectives (Kasemeier-Kulesa, Mclennan, Romine, Kulesa, & Lefcort, 2010).

Single cell migration is prominent in the Trunk Neural Crest (TNC) of the dorsolateral pathway (Fig. 2). Indeed, dorsolaterally migrating TNC cells move as single cells in the restricted spaces between the somites and ectoderm. Across vertebrate models, dorsolaterally migrating TNC appear to migrate independently of one another as single cells (Raible et al., 1992; Sadaghiani & Thiebaud, 1987; Serbedzija et al., 1989; Serbedzija et al., 1990). In zebrafish, live imaging of dorsolaterally migrating TNCs suggests they might use ameboid migration strategies (Raible et al., 1992). However, the cellular dynamics of dorsolateral TNC migration have surprisingly not been yet characterized in detail.

7. Patterning of trunk NC migration and interaction with surrounding environment

Neural crest (NC) cells migrate in versatile microenvironments. The surrounding microenvironment along the anterior-posterior axis provides distinct cues for NC cells to proceed in migration. Unlike in the cranial region, where NC migrates collectively in cell free spaces filled with extracellular components, in the trunk region NC cells are exposed to limited space between adjacent tissues. How the different physiological tissue environment impacts trunk neural crest (TNC) migration has been largely studied in avian, amphibian and mammalian embryos, while little is known about the cellular mechanisms regulating TNC migration in zebrafish embryos.

7.1 Trunk neural crest interactions with somites

The difference in the trajectories and the patterning of the stream is likely due to differences in the tissue environment between the species. In mouse and chick embryos, sclerotome and dermomyotome tissues provide repulsive cues that guide NC to take a certain trajectory; ephrin-B transmembrane ligands are expressed on the posterior region of the sclerotome while NC express receptors for those repulsive ligands (Fig. 3) (Krull et al., 1997; Theveneau & Mayor, 2012; Wang & Anderson, 1997). Alteration of ephrin-B pathway results in a loss of segmented migratory pattern and allows NC migration through the posterior sclerotome (Krull et al., 1997). In addition, in mice, NC expresses Npn2 receptor, while the expression of its repulsive ligand semapphorin3F is limited to the posterior region of the sclerotome. Inhibition of Sema3F/Nrp2 signaling leads to the loss of the segmental migration pattern of NC although this does not prevent the normal segmentation of the dorsal root ganglia (Gammill, Gonzalez, Gu, & Bronner-Fraser, 2006). Furthermore, semaphorin3A is expressed both in the posterior region of the sclerotome and dermomyotome, limiting the migration on TNC (Gammill et al., 2006). Inhibition of Sema3A/Nrp1 pathway prevents segmental migration and leads to ectopic neuronal differentiation (Schwarz, Maden, Davidson, & Ruhrberg, 2009). Inhibition of both Sema3F and Sema3A leads to ectopic invasion of the intersomitic furrows and posterior regions of the sclerotomes, causing a fusion of dorsal root ganglia (Schwarz et al., 2009). Other repulsive signals secreted by the dermomyotome in chick and mice include Slit and Draxin. Robo acts as a receptor for Slit ligands and is expressed in TNC, preventing these cells from entering the dorsolateral pathway only available for future pigment cells (Jia, Cheng, & Raper, 2005). Slit also prevents TNC from entering the developing gut (De Bellard, Rao, & Bronner-Fraser, 2003; Zuhdi et al., 2015). Draxin, a soluble protein, is found in the dorsal lip of the dermomyotome and acts as another repulsive chemokine for TNC in mouse and chick embryos (Su et al., 2009; Zhang, Su, Gao, Zhang, & Tanaka, 2017). While NC destined to take the medial pathway express receptors for repellents preventing the migration towards the dorsolateral

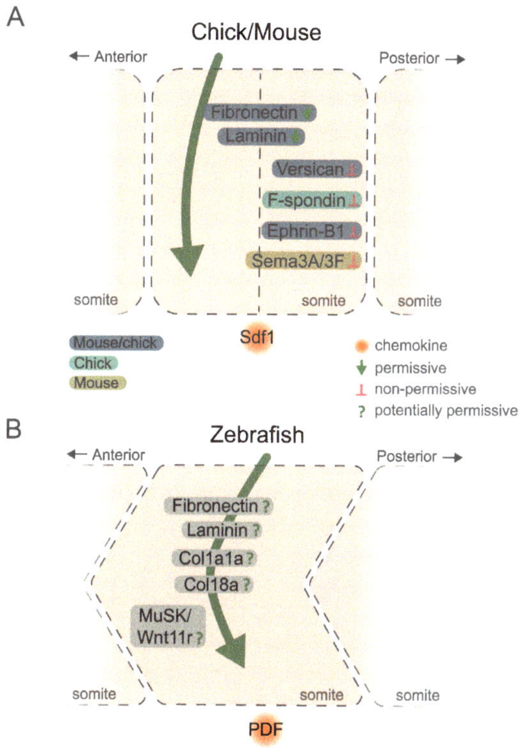

Fig. 3 Permissive and non-permissive cues form the surrounding tissues guiding the trunk NC migration in chick and mouse embryos and in zebrafish embryos. (A) Lateral view of the trunk NC migrating through the anterior half of the sclerotome. Permissive components such as fibronectin, laminin and secreted Sdf1 promotes NC migration. Non-permissive components Versican and F-spondin (in chick) as well as semaphorin (in mouse) and ephrin families are present on the posterior sclerotome restricting the NC migration to the anterior half of the sclerotome. (B) Lateral view of the trunk NC migrating on the middle region of each somite in zebrafish embryo. Permissive extracellular components such as fibronectin, laminin and Collagen1a1a, Collagen18a1 are likely permissive cues for NC migration. PDFG and MuSK/Wnt11r signaling pathways are also proposed to promote trunk NC migration in zebrafish embryo. Large green arrow indicating the direction of trunk NC migration.

path, NC restricted to the melanocytic lineage reacts positively to the same cues (Santiago & Erickson, 2002). In chick embryos, ephrin-B1, which prevents TNC from migrating on the posterior region of the sclerotome, stimulates the migration of the future pigment cells towards the lateral path (Santiago & Erickson, 2002). Indeed, melanoblast fated NC cells express the EphB2 receptor, which, in contrast to the EphB3 involved in contact repulsion responses promotes cell migration (Harris, Hall, & Erickson, 2008).

In mice, the secreted factor Steel is expressed by the dermatome and its receptor c-kit is required in dorsolateral cells for invasion and survival in the dorsolateral pathway (Wehrle-Haller & Weston, 1995; Wehrle-Haller & Weston, 1999; Wehrle-Haller, Morrison-Graham, & Weston, 1996). However, this appears to be specific to the mouse as, in chick, Steel is secreted by the ectoderm only at much later stages (Reedy, Faraco, & Erickson, 1998; Reedy, Johnson, & Erickson, 2003) and knockdown of c-kit does not affect melanoblast migration (Harris et al., 2008).

These works together support that the expression pattern of ligands along the anterior-posterior axis of the sclerotome and in dermomyotome tissues is crucial for the correct timing and patterning of TNC migration.

Whether a similar repulsive mechanism guides the TNC cell migration in zebrafish is not well understood. Live imaging of zebrafish TNC cell migration has shown a collapse of protrusions extended to the dorsolateral surface of the somite (Jesuthasan, 1996), suggesting that a repulsive mechanism could promote the initial medial path migration of zebrafish TNC cells. Unlike in mammals and avians, in teleost embryos, most of the somite-derived cells are myotome, and the sclerotome represents only a small fraction in the ventromedial aspect of the somite. Removal of the sclerotome in zebrafish does not prevent the formation of the dorsal root ganglia (Morin-Kensicki & Eisen, 1997). Instead, patterned NC migration in zebrafish trunk requires the presence of the somites, more precisely, the presence of slow-muscle precursors called adaxial cells. Reduction, and prevention of the proper segmentation or removal of the somite tissue leads to defects in NC migration (Honjo & Eisen, 2005). Defects in adaxial cells were enough to lead to defects in NC migration, although the somites were otherwise intact, and transplantation of adaxial cells was able to rescue neural crest migration in mutants lacking slow muscle cells (Honjo & Eisen, 2005).

How adaxial cells guide TNC migration is not fully understood. Muscle-specific kinase (MuSK) is crucial to maintaining TNC migration in the middle region of the somite in zebrafish (Fig. 3)(Banerjee et al., 2011). Inhibition of MuSK or its putative ligand Wnt11r leads to migration of TNC in an un-segmented manner. A similar phenotype was observed in a mutation of Wnt/MuSK downstream effector Disheveled specifically in adaxial cells (Banerjee et al., 2011), suggesting an adaxial cell dependent mechanism for patterned TNC migration in zebrafish. Zebrafish adaxial cells may contain patterning information comparable to the mammalian and avian sclerotome.

In addition to the repellent cues, secretion of chemoattractants is crucial for the directional and persistent migration of NC cells. In the trunk in mouse and chick embryo, Sdf1 soluble chemokine attracts the CXCR4 expressing NC to the ventromedial path (Belmadani et al., 2005; Kasemeier-Kulesa et al., 2010). The growth factor PDGF, which has been shown to promote chemotaxis in CNC (Bahm, et al., 2017), is important for zebrafish TNC migration (Damm & Clements, 2017). Genetic or pharmacological inhibition of PDGF receptors does not alter the initiation of TNC migration but prevents the progression after reaching the horizontal myoseptum (Damm & Clements, 2017). Furthermore, only NC in the medial path but not in the lateral path were affected, suggesting that PDGF is crucial for the persistence of migration of the cells in the medial path but other factors regulate the initiation of the migration.

7.2 Interactions with the extracellular matrix

NC migration relies on integrin based interactions with the extracellular matrix (ECM) components, including glycoproteins, collagens and proteoglycans (Perris & Perissinotto, 2000). The interactions between NC cells and the surrounding ECM are complex and dynamic and likely vary between species as well as spatiotemporally within the embryo. NC interactions with ECM can be categorized as permissive (supporting strong adhesions), non-permissive (weak/no adhesions) and inhibiting interactions (Perris & Perissinotto, 2000; Perris, 1997). The nature of the interactions varies between species, but in general, ECM components with permissive interactions include glycoproteins fibronectin and laminin, and some collagen forms (especially type I and IV) and the absence or inhibition of the interaction of NC with these components leads to defects in NC migration (Bronner-Fraser & Lallier, 1988; Perris, 1997; Perris & Perissinotto, 2000; Poole & Thiery, 1986). Non-permissive or potentially inhibitory ECM components include for example F-spondin, an ECM binding protein, and Versicans, proteoglycans, all expressed on the posterior half of the sclerotome (Fig. 3) (Debby-Brafman, Burstyn-Cohen, Klar, & Kalcheim, 1999; Dutt, Kléber, Matasci, Sommer, & Zimmermann, 2006; Landolt, Vaughan, Winterhalter, & Zimmermann, 1995). Neutralization of the F-spondin leads to ectopic migration of TNC over the posterior sclerotomes suggesting a non-permissive nature of F-spondin for NC cells (Debby-Brafman et al., 1999). Similarly, versicans have been shown to have a non-permissive or inhibitory role in NC cell migration *in vitro* (Dutt et al., 2006; Perris et al., 1996), and loss of function experiments of versican *in vivo* shows ectopic NC migration (Szabó et al., 2016). Versicans expressed in the regions surrounding NC migration have been suggested to guide NC into narrow streams promoting contact inhibition of locomotion and directional migration (Szabó et al., 2016). Another member of the proteoglycans is aggrecan, which has also been shown to have a non-permissive nature for NC migration (Perissinotto et al., 2000; Perris & Perissinotto, 2000).

In addition to glycoproteins, collagens, proteoglycans and ECM binding proteins, also glycosaminoglycans such as hyaluronan are a crucial component of ECM (Laurent & Fraser, 1992; Toole, 2001). Hyaluronan is a long un-branched polymer with a highly hydrophilic nature, allowing large hydrodynamic volume (Toole, 2001). Its long polymers allow long-range cell-surface interactions and provide a scaffold for multiple interactions, including versican and aggrecan (Perris & Perissinotto, 2000; Toole, 2001; Zimmerman, Geiger, & Addadi, 2002). However, the capacity of hyaluronan to form viscous hydrogels can also prevent more stable integrin based interactions from forming and thus can reduce cell migration (Irie et al., 2021; Zimmerman et al., 2002). Hyaluronan is expressed in the cranial and trunk region during NC migration (Nardini et al., 2004). Downregulation of hyaluronan synthesis or its receptor CD44 leads to defects in NC migration both in the cranial and trunk region (Casini, Nardi, & Ori, 2012; Ori, Nardini, Casini, Perris, & Nardi, 2006). In mice, ablation of Tmem2, a hyaluronidase that degrades high-molecular weight hyaluronan, leads to a defect in NC migration out of the neural tube (Inubushi et al., 2022). The authors proposed that the hyaluronan needs to be locally degraded to allow robust integrin based focal adhesion formation (Inubushi et al., 2022). Alternatively, local degradation of hyaluronan by Tmem2 might be important to allow correct presentation of guidance factors to the migrating neural crest cells, as this enzyme is required for VEGF signaling during collective migration of zebrafish endothelial cells (De Angelis et al., 2017a, 2017b).

Permissive or non-permissive ECM components are not well characterized in the context of zebrafish NC cell migration. Both fibronectin and laminin are expressed during zebrafish somitogenesis at somite boundaries (Crawford, Henry, Clason, Becker, & Hille, 2003; Jessen, 2015; Jülich, Geisler, Holley, & Tübingen 2000 Screen Consortium, 2005). Especially fibronectin is also localized between neural tube and somites at the onset of the TNC migration and is suggested to be relevant for intertissue adhesion during body axis elongation (Araya, Carmona-Fontaine, & Clarke, 2016;

Dray et al., 2013; Guillon et al., 2020). *In vitro,* zebrafish cranial NC migrates on collagen-I and laminin/fibronectin coated dishes, but not on laminin alone (Kinikoglu, Kong, & Liao, 2014a; Kinikoglu, Kong, & Liao, 2014b), suggesting that in zebrafish trunk, NC migration is likely based on permissive interactions with fibronectin, collagen and potentially laminin. Notably, the zebrafish genome encodes several collagen isoforms, from which some expressed in the somites, including col1a1a, col5a1, col18a1 col15B (Fang, Adams, Mcmahan, Brown, & Oxford, 2010; Thisse et al., 2001; Thisse et al., 2004; Banerjee et al., 2013; Lencer, Rains, Binne, Prekeris, & Artinger, 2023; Guillon, Bretaud, & Ruggiero, 2016) and many in notochord at the onset and during the NC migration, including col5a3, col7a1I, col9a1a, col9a2, col9a3, col11a1a, col11a2, col15a1a, col27a1a, col27a1b (Christiansen, Lang, Pace, & Parichy, 2009; Fang et al., 2010; Thisse et al., 2001). However, only few of the collagen loss-of -functions phenotypes have been characterized so far. *In vivo* work has shown that knock-down of adaxial cells deposited collagen18A1 stall TNC migration suggesting col18A1 to be permissive and required for zebrafish TNC migration (Banerjee, Isaacman-Beck, Schneider, & Granato, 2013). In addition, collagen1a1a is expressed in the somites at the onset of TNC migration, and it has been suggested to be relevant for patterned and directional TNC stream formation (Lencer et al., 2023). Since many of the loss of function experiments of different ECM components cause phenotypes also in somite formation and segmentation, it is difficult to directly address the relevance of ECM-NC interactions for patterned, persistent and directional NC migration in zebrafish. The role of another glycoprotein, tenascin-c, for NC migration remains unclear both in zebrafish and other vertebrates (Perris & Perissinotto, 2000). In zebrafish, tenascin-c is expressed in the cranial neural crest and adaxial cells in the trunk, making it an attractive candidate to participate in NC migration (Schweitzer et al., 2005; Tongiorgi, Bernhardt, Zinn, & Schachner, 1995). While tenascin-c is required for motor axon growth in zebrafish (Schweitzer et al., 2005), it is not clear whether it plays a role in NC migration. Also, hyaluronan is expressed in the intercellular space between neural tube and somites in zebrafish trunk during TNC migration (Bakkers et al., 2004; De Angelis et al., 2017a, 2017b), but its role for NC delamination or migration has not been addressed yet.

In addition to the presence of ECM components, their dynamic modification is also crucial for NC migration (Leonard & Taneyhill, 2020). Before migration, NC cells delaminate from the neural tube and undergo Epithelial-to-Mesenchymal Transition (EMT). The EMT followed by cell migration has been shown to require modification of the basal lamina to allow cell motility. Two families of proteases have been identified as key regulators of NC cell motility: Matrix Metalloproteinases (MMPs) and Disintegrin and Metalloproteinases (ADAMs). Many of these family members are expressed by NC shortly before or during the EMT and cell migration. MMPS and ADAMs can regulate EMT and NC cell migration either through modification of EMT or by regulating cadherin dependent cell-cell adhesions (Abbruzzese, Becker, Kashef, & Alfandari, 2016; Leonard & Taneyhill, 2020; Roth, Kalev-Altman, Monsonego-Ornan, & Sela-Donenfeld, 2017). In chick embryos, inhibition of MMP9 leads to defects in NC EMT and migration both at the cranial and trunk regions (Monsonego-Ornan et al., 2012). The increase or MMP9 levels correlated with decreased basal lamina at the dorsal neural tube before EMT and NC migration. Similarly, increased MMP16 correlates with increased laminin-free areas in chick embryos at the cranial region (Roth et al., 2017). These results suggest that MMP9 and MMP16 participate in basal laminin degradation required for NC EMT and migration in chick (Monsonego-Ornan et al., 2012; Roth et al., 2017). In Xenopus, MMP14 is expressed in premigratory and migratory NC cells and gain-of-function and loss-of-function *in vivo* experiments showed that MMP14 enhances NC migration in the cranial region (Garmon, Wittling, & Nie, 2018). Also, in zebrafish MMPs have been shown to participate in regulating NC migration. MMP17b is expressed in the migratory TNC, and reduction of the MMP17b leads to a defect in stream formation (Leigh et al., 2013). In addition to MMPs and ADAMs, also other players, such as glycosyltransferases can modify ECM components. In zebrafish, knockdown of the glycosyltransferase Lh3 leads to a defect in the patterning of TNC migration (Banerjee et al., 2013), leading the authors to speculate that the enzymatic modification of the non-fibrillar col18A1 through Lh3 is required for TNC stream formation in zebrafish. These examples suggest that in addition to the presence of ECM components, their dynamic remodeling is crucial for persistent and patterned NC cell migration.

7.3 Physical cues and NC migration

Mechanical properties of the microenvironment can also be a key regulator of cell migration. Recent evidence from *Xenopus* embryos shows that also NC is responsive to mechanical cues from their physiological tissue environment. *Xenopus* cranial NC have shown that expression of repellent Versican in the microenvironment creates a pattern that guides NC into narrow streams, functioning as a physical barrier promoting contact inhibition of locomotion and directional migration (Szabó et al., 2016). Also, changes in stiffness in the microenvironment have been shown to trigger NC migration in *Xenopus* cranial region; at the onset of the NC migration the underlying mesoderm increases stiffness, initiating the NC EMT and triggering the cell migration (Barriga et al., 2018). In turn, EMT results in the softening of the NC cell clusters, which is required for efficient cell migration (Marchant et al., 2022). In addition, *ex vivo* experiments using enteric NCCs from mice and chicks suggest that the behavior of these cells responds to changes in substrate stiffness (Chevalier et al., 2016).

Whether NC cell migration in the trunk responds to mechanical signals from the surrounding tissue environment has not yet been studied in detail. In the trunk, NC migrates in physically limited spaces forming narrow streams between two closely apposed tissues; the medial path between the neural tube and somites (in mice and chick between neural tube and early somite, and later between sclerotome and dermomyotome) and the dorsolateral path between ectoderm and somite (in mouse and chick between ectoderm and dermomyotome). Whether the physical proximity of the adjacent tissues regulates TNC migration has not yet been addressed. Previous studies have shown that *in vitro*, chick NC cells are capable of squeezing into pores with a diameter of 2–3 μm (Newgreen, 1989), suggesting that NC can migrate in extremely narrow spaces. However, how the mechanical forces from the surrounding ECM affect NC migration has not been yet directly addressed. The intercellular space NC is occupying is filled with ECM components, that not only provide a surface for efficient migration but also create a physical barrier that cells need to pass. Whether the navigation through small pores of fibrillar collagen mesh or viscous hyaluronan hydrogel triggers mechanosensitive responses in NC during migration remains to be explored.

8. In vitro models of neural crest: advantages and limitations

While studying neural crest (NC) cells *in vivo* is ideal, it is often limited by technical and ethical constraints. As a result, *in vitro* and *ex vivo* systems have been developed to explore NC biology. These models provide valuable platforms for testing the roles of intrinsic and extrinsic factors in NC differentiation under both normal and pathogenic conditions. Key approaches include explants, primary cell culture, and the induction of NC cells from human pluripotent stem cells (hPSCs).

Explants have proven crucial for studying NC biology, especially in species like mammals and amphibians, which lack optical clarity. Chick neural tube explants, for example, have been used to investigate NC cell morphology, behavior, adhesion, and migration (Delannet & Duband, 1992; Duband, Nekooie-Marnany, & Dufour, 2020; Loring & Erickson, 1987; Newgreen, Ritterman, & Peters, 1979; Testaz, Delannet, & Duband, 1999). To establish purer populations of cranial or trunk NC cells, chick explants can be dissociated, labeled with the NC-specific epitope HNK1, and purified using antibody-conjugated magnetic beads (Lee et al., 2020). In *Xenopus*, animal caps have been cultured *in vitro* to study NC induction (Mayor, Morgan, & Sargent, 1995). Before their migration, cranial NC cells form a distinct group that can be dissected, cultured, and imaged. Cultures of *Xenopus* CNC explants have been widely adopted (Barriga, Shellard, & Mayor, 2019) and they have allowed to discover much of the cellular and molecular principles underlying NC collective migration and EMT(Gouignard, Rouvière, & Theveneau, 2020) as they are suited for cell migration, cell dispersion and chemotaxis assays. In contrast, whilst *Xenopus* TNC cultures have been previously reported (Milos & Wilson, 1986; Milos, Wilson, Ma, Mohanraj, & Frunchak, 1987; Wilson & Milos, 1987), they are not widely used as an *in vitro* model, possibly due to the difficulty in dissecting sufficient material for functional assays. Molecular mechanisms of NC migration have also been explored using inhibitors on murine and *Xenopus* explants and graft cultures, revealing the role of glycogen synthase kinase 3 in NC regulation (Gonzalez Malagon et al., 2018).

Primary NC cultures have been developed in various model organisms. For example, Kinikoglu et al. established a protocol to dissociate and culture cranial NC cells from zebrafish embryos at the 10-somite stage. Although these cultures are heterogeneous—containing NC cells intermingled with other cell types—this approach allows for chemical screening to identify important pathways in NC development. One such screen revealed the significance of Akt activation in NC biology (Ciarlo & Zon, 2016; Ciarlo et al., 2017). For more homogeneous NC cultures, zebrafish embryos expressing EGFP under the NC-specific Sox10 promoter can be dissociated and fluorescence-activated cell sorting (FACS) used to isolate pure NC cells (Kinikoglu et al., 2014a, 2014b). Similarly, Replogle et al. generated cranial and trunk NC cultures from the same murine embryo, enabling direct comparison between these populations during differentiation. Using embryos expressing EYFP in NC cells, they were dissected at the cranial and trunk levels, dissociated, FACS-sorted, and cultured

(Replogle et al., 2018). A stable cranial NC line, O9–1, derived from mouse embryos, has also been developed. This line can be propagated, passaged, and differentiated into cranial NC derivatives (Ishii et al., 2012).

The development of 3D culture techniques led to the introduction of "crestospheres," spheroid structures formed by self-organizing NC progenitors. Crestospheres enable long-term culture of NC cells in a three-dimensional environment that better mimics the native embryonic conditions. Kerosuo et al. used crestospheres derived from human NC cells to explore stemness and differentiation potential into neuronal, mesenchymal, and melanocytic lineages (Kerosuo, Nie, Bajpai, & Bronner, 2015; Mohlin et al., 2019).

Another significant advancement in *in vitro* NC models is the conversion of other cell types into NC. For example, Motohashi et al. directly reprogrammed murine embryonic fibroblasts into trunk NC by expressing the transcription factors C-MYC, KLF4, and SOX10. These reprogrammed cells could be cultured to study NC differentiation (Motohashi et al., 2016). Induced NC cells also serve as platforms for studying NC-related diseases, such as neurocristopathies. For instance, forced Sox10 expression in postnatal human fibroblasts from healthy and familial dysautonomia patients generated induced NC cells, with CD34 identified as a marker for their selection. These cells exhibited functional, multipotent properties and disease-relevant phenotypes (Kim et al., 2014). Another strategy involved converting human inter-follicular keratinocytes from elderly donors into NC cells using a growth factor cocktail containing FGF2 and IGF1. These cells were able to migrate normally in chick embryos and differentiate into all NC derivatives, offering a potential source for disease modeling and drug discovery (Moghadasi Boroujeni et al., 2019).

Human pluripotent stem cells (hPSCs) have also been employed to induce NC cells for research. In one study, hPSCs were used to create 3D patterned branchial arch aggregates by adding BMP4 and EDN1, allowing the study of NC specification and patterning during early facial development (Seto, Ogihara, Takizawa, & Eiraku, 2024). Fan et al. demonstrated that hPSCs could be converted into various NC subtypes, including vagal and sacral NC, which contribute to the enteric nervous system (ENS). These cells were used to rescue a Hirschsprung's disease mouse model, highlighting therapeutic potential. High levels of WNT and FGF induced a neuromesodermal progenitor, with subsequent GDF11 addition producing sacral NC, while the absence of GDF11 resulted in trunk NC cells. Cranial NC was generated by low WNT levels while the addition of retinoic acid, induced vagal NC identity (Fan et al., 2023; Fattahi et al., 2016; Huang et al., 2016). Another protocol efficiently differentiated hPSCs into cranial or trunk NC by manipulating WNT and FGF signaling, driving the expression of characteristic HOX genes (Frith et al., 2018; Hackland et al., 2019).

Workman et al. generated NC-derived organoids from induced pluripotent stem cells, which contained vagal NC derivatives, including enteric neurons and glia. This model recapitulated key features of ENS development and was used to model Hirschsprung's disease, which is caused by defective NC migration leading to a lack of enteric ganglia (Workman et al., 2016).

In vitro models have provided crucial insights into NC biology, particularly in understanding NC migration, differentiation, and underlying molecular mechanisms. These systems, such as explants, primary cultures, and 3D models, allow for controlled experiments to study NC behavior and pathologies, as well as develop therapeutic strategies. However, *in vitro* models, especially in 2D cultures, lack the complexity of the *in vivo* environment, which may influence NC behavior.

In summary, *in vitro* systems have greatly advanced our understanding of NC development, revealing critical pathways and mechanisms that govern their contributions to vertebrate development and disease.

9. Conclusion

The study of TNC offers valuable insights into the complex mechanisms of collective cell migration, cell migration in complex environments and cell fate specification in development. TNCCs navigate their migratory routes and differentiate into various specialized cell types following signaling pathways and complex environments. While significant progress has been made in understanding the mechanisms underlying fate acquisition, collective migration and environmental signals guiding TNC routes, many questions remain regarding the precise timing and nature of these signals *in vivo*. As research continues to uncover the nuances of NCC development, we can expect to gain a deeper appreciation of both vertebrate biology and the underlying mechanisms that contribute to developmental disorders.

Understanding these processes not only advances our knowledge of vertebrate evolution but also has potential implications for regenerative medicine and the treatment of hereditary diseases. The dynamic nature of NCCs, coupled with their ability to respond to environmental cues, highlights the significant adaptability of these cells during development, making them a fascinating subject of study in the field of developmental biology.

References

Abbruzzese, G., Becker, S. F., Kashef, J., & Alfandari, D. (2016). ADAM13 cleavage of cadherin-11 promotes CNC migration independently of the homophilic binding site. *Developmental Biology, 415*(2), 383–390. https://doi.org/10.1016/J.Ydbio.2015.07.018.

Abercrombie, M., & Heaysman, J. E. (1953). Observations on the social behaviour of cells in tissue culture. I. Speed of movement of chick heart fibroblasts in relation to their mutual contacts. *Experimental Cell Research, 5*, 111–131.

Akitaya, T., & Bronner-Fraser, M. (1992). Expression of cell adhesion molecules during initiation and cessation of neural crest cell migration. *Developmental Dynamics: An Official Publication of the American Association of Anatomists, 194*, 12–20.

Alhashem, Z., Camargo-Sosa, K., Kelsh, R. N., & Linker, C. (2022). Trunk neural crest migratory position and asymmetric division predict terminal differentiation. *Frontiers in Cell and Developmental Biology, 10*, 887393.

Alhashem, Z., Feldner-Busztin, D., Revell, C., Alvarez-Garcillan Portillo, M., Camargo-Sosa, K., Richardson, J., ... Linker, C. (2022). Notch controls the cell cycle to define leader versus follower identities during collective cell migration. *Elife,* 11.

Anderson, D. J. (1989). The neural crest cell lineage problem. *Neuropoiesis? Neuron, 3*(1), 1–12. https://doi.org/10.1016/0896-6273(89)90110-4.

Araya, C., Carmona-Fontaine, C., & Clarke, J. D. (2016). Extracellular matrix couples the convergence movements of mesoderm and neural plate during the early stages of neurulation. *Developmental Dynamics: An Official Publication Of The American Association Of Anatomists, 245*(5), 580–589. https://doi.org/10.1002/Dvdy.24401.

Baggiolini, A., Varum, S., Mateos, J. M., Bettosini, D., John, N., Bonalli, M., ... Sommer, L. (2015). Premigratory and migratory neural crest cells are multipotent in vivo. *Cell Stem Cell, 16*, 314–322.

Bahm, I., Barriga, Eh, Frolov, A., Theveneau, E., Frankel, P., & Mayor, R. (2017). Pdgf controls contact inhibition of locomotion by regulating N-cadherin during neural crest migration. *Development (Cambridge, England), 144*(13), 2456–2468. https://doi.org/10.1242/Dev.147926 Epub 2017 May 19. Pmid: 28526750; Pmcid: Pmc5536867.

Bakkers, J., Kramer, C., Pothof, J., Quaedvlieg, N. E., Spaink, H. P., & Hammerschmidt, M. (2004). Has2 is required upstream of Rac1 to govern dorsal migration of lateral cells during zebrafish gastrulation. *Development (Cambridge, England), 131*(3), 525–537. https://doi.org/10.1242/Dev.00954.

Banerjee, S., Gordon, L., Donn, T. M., Berti, C., Moens, C. B., Burden, S. J., & Granato, M. (2011). A novel role for MuSK and non-canonical Wnt signaling during segmental neural crest cell migration. *Development, 138*, 3287–3296. https://doi.org/10.1242/dev.067306.

Banerjee, S., Isaacman-Beck, J., Schneider, V. A., & Granato, M. (2013). A novel role for Lh3 dependent Ecm modifications during neural crest cell migration in zebrafish. *PLoS One, 8*(1), E54609. https://doi.org/10.1371/Journal.Pone.0054609.

Barriga, E. H., Franze, K., Charras, G., & Mayor, R. (2018). Tissue stiffening coordinates morphogenesis by triggering collective cell migration in vivo. *Nature, 554*, 523–527.

Barriga, E. H., Maxwell, P. H., Reyes, A. E., & Mayor, R. (2013). The hypoxia factor Hif-1alpha controls neural crest chemotaxis and epithelial to mesenchymal transition. *The Journal of Cell Biology, 201*, 759–776.

Barriga, Eh, Shellard, A., & Mayor, R. (2019). In vivo and in vitro quantitative analysis of neural crest cell migration. In Q. Schwarz, & S. Wiszniak (Eds.). *Neural crest cells: Methods and protocols* (pp. 135–152). New York: Springer. 10.1007/978-1-4939-9412-0_11.

Becker, S. F., Mayor, R., & Kashef, J. (2013). Cadherin-11 mediates contact inhibition of locomotion during xenopus neural crest cell migration. *PLoS One, 8*, E85717.

Belmadani, A., Tran, P. B, Ren, D., Assimacopoulos, S., Grove, E. A, & Miller, R. J (2005). The chemokine stromal cell-derived factor-1 regulates the migration of sensory neuron progenitors. *The Journal of Neuroscience, 25*(16), 3995–4003. https://doi.org/10.1523/Jneurosci.4631-04.2005 Pmid: 15843601; Pmcid: Pmc4461238.

Betancur, P., Bronner-Fraser, M., & Sauka-Spengler, T. (2010). Assembling neural crest regulatory circuits into a gene regulatory network. *Annual Review of Cell and Developmental Biology, 26*, 581–603.

Breau, M. A., Pietri, T., Stemmler, M. P., Thiery, J. P., & Weston, J. A. (2008). A nonneural epithelial domain of embryonic cranial neural folds gives rise to ectomesenchyme. *Proceedings of the National Academy of Sciences of the United States of America, 105*, 7750–7755.

Bronner-Fraser, M., & Fraser, S. E. (1988). Cell lineage analysis reveals multipotency of some avian neural crest cells. *Nature, 335*(6186), 161–164. https://doi.org/10.1038/335161a0.

Bronner-Fraser, M., & Fraser, S. (1989). Developmental potential of avian trunk neural crest cells in situ. *Neuron, 3*(6), 755–766. https://doi.org/10.1016/0896-6273(89)90244-4.

Bronner-Fraser, M., & Lallier, T. (1988). A monoclonal antibody against a laminin-heparan sulfate proteoglycan complex perturbs cranial neural crest migration in vivo. *The Journal of Cell Biology, 106*(4), 1321–1329. https://doi.org/10.1083/Jcb.106.4.1321.

Carmona-Fontaine, C., Matthews, H. K., Kuriyama, S., Moreno, M., Dunn, G. A., Parsons, M., ... Mayor, R. (2008). Contact inhibition of locomotion in vivo controls neural crest directional migration. *Nature, 456*, 957–961.

Carmona-Fontaine, C., Theveneau, E., Tzekou, A., Tada, M., Woods, M., Page, K. M., ... Mayor, R. (2011). Complement fragment C3a controls mutual cell attraction during collective cell migration. *Developmental Cell, 21*, 1026–1037. https://doi.org/10.1016/j.devcel.2011.10.012.

Casini, P., Nardi, I., & Ori, M. (2012). Hyaluronan is required for cranial neural crest cells migration and craniofacial development. *Developmental Dynamics: An Official Publication of the American Association of Anatomists, 241*(2), 294–302. https://doi.org/10.1002/Dvdy.23715.

Chacon-Martinez, C. A., Koester, J., & Wickstrom, S. A. (2018). Signaling in the stem cell niche: Regulating cell fate, function and plasticity. *Development (Cambridge, England), 145*.

Chalpe, A. J., Prasad, M., Henke, A. J., & Paulson, A. F. (2010). Regulation of cadherin expression in the chicken neural crest by the Wnt/beta-catenin signaling pathway. *Cell Adhesion & Migration, 4*, 431–438.

Cheung, M., Chaboissier, M. C., Mynett, A., Hirst, E., Schedl, A., & Briscoe, J. (2005). The transcriptional control of trunk neural crest induction, survival, and delamination. *Developmental Cell, 8*, 179–192.

Chevalier, N. R., Gazquez, E., Bidault, L., Guilbert, T., Vias, C., Vian, E., ... Fleury, V. (2016). Corrigendum: How tissue mechanical properties affect enteric neural crest cell migration. *Scientific Reports, 6*, 34118. https://doi.org/10.1038/Srep34118.

Chibon, P. (1964). Analysis by the method of nuclear labelling with tritiated thymidine of derivatives of the cephalic neural crest in the urodele Pleurodeles waltlii michah. *Comptes Rendus Hebdomadaires des Seances de l'Academie des Sciences, 259*, 3624–3627.

Chibon, P. (1967). Nuclear labelling by tritiated thymidine of neural crest derivatives in the amphibian urodele Pleurodeles waltlii michah. *Journal of Embryology and Experimental Morphology, 18*, 343–358.

Christiansen, H. E., Lang, M. R., Pace, J. M., & Parichy, D. M. (2009). Critical early roles for Col27a1a and Col27a1b in zebrafish notochord morphogenesis, vertebral mineralization and post-embryonic axial growth. *PLoS One, 4*(12), E8481. https://doi.org/10.1371/Journal.Pone.0008481.

Ciarlo, C., Kaufman, C. K., Kinikoglu, B., Michael, J., Yang, S., D'Amato, C., ... Zon, Li (2017). A chemical screen in zebrafish embryonic cells establishes that akt activation is required for neural crest development. *Elife, 6*, E29145. https://doi.org/10.7554/Elife.29145.

Ciarlo, Ca, & Zon, Li (2016). Embryonic cell culture in zebrafish. In Hw Detrich, M. Westerfield, & I. Zon Leonard (Eds.). *Methods in cell biology, the zebrafish* (pp. 1–10). Academic Press. https://doi.org/10.1016/Bs.Mcb.2016.02.010.

Clay, M. R., & Halloran, M. C. (2014). Cadherin 6 promotes neural crest cell detachment via F-actin regulation and influences active Rho distribution during epithelial-to-mesenchymal transition. *Development (Cambridge, England), 141*, 2506–2515.

Coles, E. G., Taneyhill, L. A., & Bronner-Fraser, M. (2007). A critical role for cadherin6b in regulating avian neural crest emigration. *Developmental Biology, 312*, 533–544.

Collazo, A., Bronner-Fraser, M., & Fraser, S. E. (1993). Vital dye labelling of Xenopus laevis trunk neural crest reveals multipotency and novel pathways of migration. *Development (Cambridge, England), 118*(2), 363–376. https://doi.org/10.1242/Dev.118.2.363.

Cornell, R. A., & Eisen, J. S. (2005). Notch in the pathway: The roles of Notch signaling in neural crest development. *Seminars in Cell & Developmental Biology, 16*(6), 663–672. https://doi.org/10.1016/j.semcdb.2005.06.009.

Crawford, B. D., Henry, C. A., Clason, T. A., Becker, A. L., & Hille, M. B. (2003). Activity and distribution of paxillin, focal adhesion kinase, and cadherin indicate cooperative roles during zebrafish morphogenesis. *Molecular Biology of the Cell, 14*(8), 3065–3081. https://doi.org/10.1091/Mbc.E02-08-0537.

Dady, A., Blavet, C., & Duband, J. L. (2012). Timing and kinetics of E- To N-cadherin switch during neurulation in the avian embryo. *Developmental Dynamics: An Official Publication of the American Association of Anatomists, 241*, 1333–1349.

Damm, E. W., & Clements, W. K. (2017). Pdgf signalling guides neural crest contribution to the haematopoietic stem cell specification niche. *Nature Cell Biology, 19*(5), 457–467. https://doi.org/10.1038/Ncb3508 Epub 2017 Apr 10. Pmid: 28394883; Pmcid: Pmc5546139.

Dawes, J. H. P., & Kelsh, R. N. (2021). Cell fate decisions in the neural crest, from pigment cell to neural development. *International Journal of Molecular Sciences, 22*(24), 13531. https://doi.org/10.3390/ijms222413531.

De Angelis, J. E., Lagendijk, A. K., Chen, H., Tromp, A., Bower, N. I., Tunny, K. A., ... Smith, K. A. (2017a). Tmem2 regulates embryonic vegf signaling by controlling hyaluronic acid turnover. *Developmental Cell, 40*, 421.

De Angelis, J. E., Lagendijk, A. K., Chen, H., Tromp, A., Bower, N. I., Tunny, K. A., ... Smith, K. A. (2017b). Tmem2 regulates embryonic Vegf signaling by controlling hyaluronic acid turnover. *Developmental Cell, 40*(2), 123–136. https://doi.org/10.1016/J.Devcel.2016.12.017.

De Bellard, M. E., Rao, Y., & Bronner-Fraser, M. (2003). Dual function of Slit2 in repulsion and enhanced migration of trunk, but not vagal, neural crest cells. *The Journal of Cell Biology, 162*(2), 269–279. https://doi.org/10.1083/Jcb.200301041 Pmid: 12876276; Pmcid: Pmc2172792.

Debby-Brafman, A., Burstyn-Cohen, T., Klar, A., & Kalcheim, C. (1999). F-spondin, expressed in somite regions avoided by neural crest cells, mediates inhibition of distinct somite domains to neural crest migration. *Neuron, 22*(3), 475–488. https://doi.org/10.1016/S0896-6273(00)80703-5.

Delannet, M., & Duband, J.-L. (1992). Transforming growth factor-β control of cell-substratum adhesion during avian neural crest cell emigration in vitro. *Development (Cambridge, England), 116*, 275–287. https://doi.org/10.1242/Dev.116.1.275.

Dray, N., Lawton, A., Nandi, A., Jülich, D., Emonet, T., & Holley, S. A. (2013). Cell-fibronectin interactions propel vertebrate trunk elongation via tissue mechanics. *Current Biology: CB, 23*(14), 1335–1341. https://doi.org/10.1016/J.Cub.2013.05.052.

Duband, J.-L., Nekooie-Marnany, N., & Dufour, S. (2020). Establishing primary cultures of trunk neural crest cells. *88*, E109. https://doi.org/10.1002/Cpcb.109.

Duband, J. L., Volberg, T., Sabanay, I., Thiery, J. P., & Geiger, B. (1988). Spatial and temporal distribution of the adherens-junction-associated adhesion molecule A-CAM during avian embryogenesis. *Development (Cambridge, England), 103*, 325–344.

Dutt, S., Kléber, M., Matasci, M., Sommer, L., & Zimmermann, D. R. (2006). Versican V0 and V1 guide migratory neural crest cells. *The Journal of Biological Chemistry, 281*(17), 12123–12131. https://doi.org/10.1074/Jbc.M510834200.

Elosegui-Artola, A., Trepat, X., & Roca-Cusachs, P. (2018). Control of mechanotransduction by molecular clutch dynamics. *Trends in Cell Biology, 28*, 356–367.

Emtenani, S., Martin, E. T., Gyoergy, A., Bicher, J., Genger, J. W., Kocher, T., ... Siekhaus, D. E. (2022). Macrophage mitochondrial bioenergetics and tissue invasion are boosted by an atossa-porthos axis in Drosophila. *The EMBO Journal, 41*, E109049.

Erickson, A. G., Kameneva, P., & Adameyko, I. (2023). The transcriptional portraits of the neural crest at the individual cell level. *Seminars in Cell & Developmental Biology, 138*, 68–80.

Ernsberger, U., Patzke, H., Tissier-Seta, J. P., Reh, T., Goridis, C., & Rohrer, H. (1995). The expression of tyrosine hydroxylase and the transcription factors cPhox-2 and Cash-1: Evidence for distinct inductive steps in the differentiation of chick sympathetic precursor cells. *Mechanisms of Development, 52*(1), 125–136. https://doi.org/10.1016/0925-4773(95)00396-I.

Etchevers, H. C., Amiel, J., & Lyonnet, S. (2007). Genetic and molecular bases of neurocristopathies. *Archives of Pediatrics, 14*, 668–672.

Fan, Y., Hackland, J., Baggiolini, A., Hung, Ly, Zhao, H., Zumbo, P., ... Studer, L. (2023). Hpsc-derived sacral neural crest enables rescue in a severe model of Hirschsprung's disease. E9 *Cell Stem Cell, 30*, 264–282. https://doi.org/10.1016/J.Stem.2023.02.003.

Fang, M., Adams, J. S., Mcmahan, B. L., Brown, R. J., & Oxford, J. T. (2010). The expression patterns of minor fibrillar collagens during development in zebrafish. *Gene Expression Patterns: GEP, 10*(7–8), 315–322. https://doi.org/10.1016/J.Gep.2010.07.002.

Fattahi, F., Steinbeck, Ja, Kriks, S., Tchieu, J., Zimmer, B., Kishinevsky, S., ... Studer, L. (2016). Deriving human ENS lineages for cell therapy and drug discovery in Hirschsprung's disease. *Nature, 531*, 105. https://doi.org/10.1038/Nature16951.

Fraser, S. E., & Bronner-Fraser, M. (1991). Migrating neural crest cells in the trunk of the avian embryo are multipotent. *Development (Cambridge, England), 112*, 913–920.

Frith, T. J., Granata, I., Wind, M., Stout, E., Thompson, O., Neumann, K., ... Tsakiridis, A. (2018). Human axial progenitors generate trunk neural crest cells in vitro. *Elife, 7*, E35786. https://doi.org/10.7554/Elife.35786.

Gammill, L. S., Gonzalez, C., & Bronner-Fraser, M. (2007). Neuropilin 2/Semaphorin 3f signaling is essential for cranial neural crest migration and trigeminal ganglion condensation. *Developmental Neurobiology, 67*, 47–56.

Gammill, L. S., Gonzalez, C., Gu, C., & Bronner-Fraser, M. (2006). Guidance of trunk neural crest migration requires neuropilin 2/Semaphorin 3f signaling. *Development (Cambridge, England), 133*(1), 99–106. https://doi.org/10.1242/Dev.02187.

Gans, C., & Northcutt, R. G. (1983). Neural crest and the origin of vertebrates: A new head. *Science (New York, N. Y.), 220*, 268–273.

Garde, A., & Sherwood, D. R. (2021). Fueling cell invasion through extracellular matrix. *Trends in Cell Biology, 31*, 445–456. https://doi.org/10.1016/j.tcb.2021.01.006.

Garmon, T., Wittling, M., & Nie, S. (2018). Mmp14 regulates cranial neural crest epithelial-to-mesenchymal transition and migration. *Developmental Dynamics: An Official Publication of the American Association of Anatomists, 247*(9), 1083–1092. https://doi.org/10.1002/Dvdy.24661.

Gonzalez Malagon, S. G., Lopez Muñoz, A. M., Doro, D., Bolger, T. G., Poon, E., Tucker, E. R., ... Liu, K. J. (2018). Glycogen synthase kinase 3 controls migration of the neural crest lineage in mouse and xenopus. *Nature Communications, 9*, 1126. https://doi.org/10.1038/S41467-018-03512-5.

Gossler, A., & Hrabe De Angelis, M. (1998). Somitogenesis. *Current Topics in Developmental Biology, 38*, 225–287.

Gouignard, N., Andrieu, C., & Theveneau, E. (2018). Neural crest delamination and migration: Looking forward to the next 150 years. *Genesis (New York, N. Y.: 2000), 56*, E23107.

Gouignard, N., Rouvière, C., & Theveneau, E. (2020). The Epithelial-to Mesenchymal Transition, Methods and Protocols. *Methods in Molecular Biology, 2179*, 257–274. https://doi.org/10.1007/978-1-0716-0779-4_20.

Guillon, E., Bretaud, S., & Ruggiero, F. (2016). Slow muscleprecursors lay down a collagen XV matrix fingerprint to guide motor axon navigation. *Journal of Neuroscience, 36*, 2663–2676. https://doi.org/10.1523/jneurosci.2847-15.2016.

Guillon, E., Das, D., Jülich, D., Hassan, A. R., Geller, H., & Holley, S. (2020). Fibronectin is a smart adhesive that both influences and responds to the mechanics of early spinal column development. *Elife, 9*, E48964. https://doi.org/10.7554/Elife.48964.

Hackland, J., Shelar, P., Sandhu, N., Prasad, M. S., Charney, R. M., Gomez, G. A, ... García-Castro, M. I (2019). FGF modulates the axial identity of trunk HPSC-derived neural crest but not the cranial-trunk decision. *Stem Cell Reports, 12*, 920–933. https://doi.org/10.1016/J.Stemcr.2019.04.015.

Hammond, W. S., & Yntema, C. L. (1947). Depletions in the thoraco-lumbar sympathetic system following removal of neural crest in the chick. *The Journal of Comparative Neurology, 86*, 237–265.

Harris, M. L., Hall, R., & Erickson, C. A. (2008). Directing pathfinding along the dorsolateral path—The role of EDNRB2 and EPHB2 in overcoming inhibition. *Development (Cambridge, England), 135*, 4113–4122.

Henion, P. D., & Weston, J. A. (1997). Timing and pattern of cell fate restrictions in the neural crest lineage. *Development (Cambridge, England), 124*, 4351–4359.

Honjo, Y., & Eisen, J. S. (2005). Slow muscle regulates the pattern of trunk neural crest migration in zebrafish. *Development (Cambridge, England), 132*(20), 4461–4470. https://doi.org/10.1242/Dev.02026.

Honjo, Y., Kniss, J., & Eisen, J. S. (2008). Neuregulin-mediated ERBB3 signaling is required for formation of zebrafish dorsal root ganglion neurons. *Development (Cambridge, England), 135*, 2615–2625.

Huang, M., Miller, Ml, Mchenry, Lk, Zheng, T., Zhen, Q., Ilkhanizadeh, S., ... Weiss, W. A. (2016). Generating trunk neural crest from human pluripotent stem cells. *Scientific Reports, 6*, 19727. https://doi.org/10.1038/Srep19727.

Hutchins, E. J., Kunttas, E., Piacentino, M. L., Howard, A. G. A. T., Bronner, M. E., & Uribe, R. A. (2018). Migration and diversification of the vagal neural crest. *Developmental Biology, 444*(1), S98–S109.

Inubushi, T., Nakanishi, Y., Abe, M., Takahata, Y., Nishimura, R., Kurosaka, H., ... Yamaguchi, Y. (2022). The cell surface hyaluronidase TMEM2 plays an essential role in mouse neural crest cell development and survival. *PLoS Genetics, 18*(7), E1009765. https://doi.org/10.1371/Journal.Pgen.1009765.

Irie, F., Tobisawa, Y., Murao, A., Yamamoto, H., Ohyama, C., & Yamaguchi, Y. (2021). The cell surface hyaluronidase TMEM2 regulates cell adhesion and migration via degradation of hyaluronan at focal adhesion sites. *The Journal of Biological Chemistry, 296*, 100481. https://doi.org/10.1016/J.Jbc.2021.100481 Epub 2021 Feb 26. Pmid: 33647313; Pmcid: Pmc8042168.

Ishii, M., Arias, A. A., Liu, L., Chen, Y.-B., Bronner, M. E., & Maxson, R. E. (2012). A stable cranial neural crest cell line from mouse. *Stem Cells and Development, 21*, 3069–3080. https://doi.org/10.1089/Scd.2012.0155.

Jessen, J. R. (2015). Recent advances in the study of zebrafish extracellular matrix proteins. *Developmental Biology, 401*(1), 110–121. https://doi.org/10.1016/J.Ydbio.2014.12.022.

Jesuthasan, S. (1996). Contact inhibition/collapse and pathfinding of neural crest cells in the zebrafish trunk. *Development (Cambridge, England), 122*(1), 381–389. https://doi.org/10.1242/Dev.122.1.381.

Jia, L., Cheng, L., & Raper, J. (2005). Slit/Robo signaling is necessary to confine early neural crest cells to the ventral migratory pathway in the trunk. *Developmental Biology, 282*(2), 411–421. https://doi.org/10.1016/J.Ydbio.2005.03.021.

Jülich, D., Geisler, R., Holley, S. A. & Tübingen 2000 Screen Consortium. (2005). Integrinalpha5 and Delta/Notch signaling have complementary spatiotemporal requirements during zebrafish somitogenesis. Developmental Cell, 8(4), 575–586. https://doi.org/10.1016/J.Devcel.2005.01.016.

Kang, J. S., Feinleib, J. L., Knox, S., Ketteringham, M. A., & Krauss, R. S. (2003). Promyogenic members of the Ig and cadherin families associate to positively regulate differentiation. *Proceedings of the National Academy of Sciences of the United States of America, 100*, 3989–3994.

Kasemeier-Kulesa, J. C., Mclennan, R., Romine, M. H., Kulesa, P. M., & Lefcort, F. (2010). Cxcr4 controls ventral migration of sympathetic precursor cells. *The Journal of Neuroscience, 30*(39), 13078–13088. https://doi.org/10.1523/Jneurosci.0892-10.2010 Pmid: 20881125; Pmcid: Pmc6633505.

Keller, R. (2000). The origin and morphogenesis of amphibian somites. *Current Topics in Developmental Biology, 47*, 183–246.

Kelsh, R. N., Camargo Sosa, K., Farjami, S., Makeev, V., Dawes, J. H. P., & Rocco, A. (2021). Cyclical fate restriction: A new view of neural crest cell fate specification. *Development (Cambridge, England), 148*.

Kerosuo, L., Nie, S., Bajpai, R., & Bronner, M. E. (2015). Crestospheres: Long-Term Maintenance of Multipotent, Premigratory Neural Crest Stem Cells. *Stem Cell Reports, 5*, 499–507. https://doi.org/10.1016/j.stemcr.2015.08.017.

Kim, Y. J., Lim, H., Li, Z., Oh, Y., Kovlyagina, I., Choi, I. Y, ... Lee, G. (2014). Generation of multipotent induced neural crest by direct reprogramming of human postnatal fibroblasts with a single transcription factor. *Cell Stem Cell, 15*, 497–506. https://doi.org/10.1016/J.Stem.2014.07.013.

Kinikoglu, B., Kong, Y., & Liao, E. C. (2014a). Featured article: Characterization of cultured multipotent zebrafish neural crest cells. *Experimental Biology and Medicine, 239*, 159–168. https://doi.org/10.1177/1535370213513997.

Kinikoglu, B., Kong, Y., & Liao, E. C. (2014b). Characterization of cultured multipotent zebrafish neural crest cells. *Experimental Biology and Medicine (Maywood, N. J.), 239*(2), 159–168. https://doi.org/10.1177/1535370213513997.

Klymkowsky, M., Cortez Rossi, C., & Artinger, K. B. (2010). Mechanisms driving neural crest induction and migration in the zebrafish and *Xenopus laevis*. *Cell Adhesion & Migration, 4*(4), 595–608. https://doi.org/10.4161/cam.4.4.12962.

Knight, R. D., & Schilling, T. F. (2006). Cranial neural crest and development of the head skeleton. *Advances in Experimental Medicine and Biology, 589*, 120–133.

Kotini, M., Barriga, E. H., Leslie, J., Gentzel, M., Rauschenberger, V., Schambony, A., & Mayor, R. (2018). Gap junction protein connexin-43 is a direct transcriptional regulator of N-cadherin in vivo. *Nature Communications, 9*, 3846.

Krotoski, D. M., & Bronner-Fraser, M. (1986). Mapping of neural crest pathways in Xenopus laevis. *Progress in Clinical and Biological Research, 217b*, 229–233.

Krull, C. E., Lansford, R., Gale, N. W., Collazo, A., Marcelle, C., Yancopoulos, G. D., ... Bronner-Fraser, M. (1997). Interactions of EPH-related receptors and ligands confer rostrocaudal pattern to trunk neural crest migration. *Current Biology: CB, 7*(8), 571–580. https://doi.org/10.1016/S0960-9822(06)00256-9.

Kuriyama, S., Theveneau, E., Benedetto, A., Parsons, M., Tanaka, M., Charras, G., ... Mayor, R. (2014). In vivo collective cell migration requires an LPAR2-dependent increase in tissue fluidity. *The Journal of Cell Biology, 206*, 113–127.

Landolt, R. M., Vaughan, L., Winterhalter, K. H., & Zimmermann, D. R. (1995). Versican is selectively expressed in embryonic tissues that act as barriers to neural crest cell migration and axon outgrowth. *Development (Cambridge, England), 121*(8), 2303–2312. https://doi.org/10.1242/Dev.121.8.2303.

Laurent, T. C., & Fraser, J. R. (1992). Hyaluronan. *FASEB Journal: Official Publication of the Federation of American Societies for Experimental Biology, 6*(7), 2397–2404.

Le Douarin, N. M. K. C. (1999). *The neural crest*. Cambridge: Cambridge University Press.

Le Douarin, N. M., & Dupin, E. (2018). The "beginnings" of the neural crest. *Developmental Biology, 444*(1), S3–S13.

Le Douarin, N. M., & Teillet, M. A. (1974). Experimental analysis of the migration and differentiation of neuroblasts of the autonomic nervous system and of neurectodermal mesenchymal derivatives, using a biological cell marking technique. *Developmental Biology, 41*, 162–184.

Le Douarin, N. M., Teillet, M. A., & Le Lievre, C. (1977). Influence of the tissue environment on the differentiation of neural crest cells. *Society of General Physiologists Series, 32*, 11–27.

Le Lievre, C., & Le Douarin, N. (1974). Ectodermic origin of the derma of the face and neck, demonstrated by interspecific combinations in the bird embryo. *Comptes rendus hebdomadaires des séances de l'Académie des sciences, 278*, 517–520.

Lee, V. M, Hernandez, S., Giang, B., Chabot, C., Hernandez, J., & De Bellard, M. E. (2020). Molecular events controlling cessation of trunk neural crest migration and onset of differentiation. *Frontiers in Cell and Developmental Biology, 8*. https://doi.org/10.3389/Fcell.2020.00199.

Leigh, N. R., Schupp, M. O., Li, K., Padmanabhan, V., Gastonguay, A., Wang, L., ... Ramchandran, R. (2013). Mmp17b is essential for proper neural crest cell migration in vivo. *PLoS One, 8*(10), E76484. https://doi.org/10.1371/Journal.Pone.0076484.

Lencer, E., Prekeris, R., & Artinger, K. B. (2021). Single-cell RNA analysis identifies premigratory neural crest cells expressing markers of differentiated derivatives. *Elife, 10*.

Lencer, E., Rains, A., Binne, E., Prekeris, R., & Artinger, K. B. (2023). Mutations in cdon and boc affect trunk neural crest cell migration and slow-twitch muscle development in zebrafish. *Development (Cambridge, England), 150*(14), Dev201304 . https://doi.org/10.1242/Dev.201304.

Leonard, C. E., & Taneyhill, L. A. (2020). The road best traveled: Neural crest migration upon the extracellular matrix. *Seminars in Cell & Developmental Biology, 100*, 177–185. https://doi.org/10.1016/J.Semcdb.2019.10.013.

Li, Y., Vieceli, F. M., Gonzalez, W. G., Li, A., Tang, W., Lois, C., & Bronner, M. E. (2019). In vivo quantitative imaging provides insights into trunk neural crest migration. *Cell Reports, 26*, 1489–1500 E3.

Loring, J. F., & Erickson, C. A. (1987). Neural crest cell migratory pathways in the trunk of the chick embryo. *Developmental Biology, 121*, 220–236. https://doi.org/10.1016/0012-1606(87)90154-0.

Lu, M., & Krauss, R. S. (2010). N-cadherin ligation, but not sonic hedgehog binding, initiates cdo-dependent P38alpha/beta mapk signaling in skeletal myoblasts. *Proceedings of the National Academy of Sciences of the United States of America, 107*, 4212–4217.

Marchant, C. L., Malmi-Kakkada, A. N., Espina, J. A., & Barriga, E. H. (2022). Cell clusters softening triggers collective cell migration in vivo. *Nature Materials, 21*, 1314–1323.

Martik, M. L., & Bronner, M. E. (2017). Regulatory logic underlying diversification of the neural crest. *Trends in Genetics: TIG, 33*, 715–727.

Mayor, R., & Etienne-Manneville, S. (2016). The front and rear of collective cell migration. *Nature Reviews. Molecular Cell Biology, 17*, 97–109.

Mayor, R., Morgan, R., & Sargent, Mg (1995). Induction of the prospective neural crest of xenopus. *Development (Cambridge, England), 121*, 767–777. https://doi.org/10.1242/Dev.121.3.767.

Mclennan, R., Teddy, J. M., Kasemeier-Kulesa, J. C., Romine, M. H., & Kulesa, P. M. (2010). Vascular endothelial growth factor (VEGF) regulates cranial neural crest migration in vivo. *Developmental Biology, 339*, 114–125.

Milos, N. C., Wilson, H. C., Ma, Y. L., Mohanraj, T. M., & Frunchak, Y. N. (1987). Studies on cellular adhesion of xenopus laevis melanophores: Modulation of cell-cell and cell-substratum adhesion in vitro by endogenous xenopus galactoside-binding lectin. *Pigment Cell Research/Sponsored by the European Society for Pigment Cell Research and the International Pigment Cell Society, 1*, 188–196.

Milos, N. C., & Wilson, H. C. (1986). Cell surface carbohydrate involvement in controlling the adhesion and morphology of neural crest cells and melanophores of Xenopus laevis. *The Journal of Experimental Zoology, 238*, 211–224.

Moghadasi Boroujeni, S., Koontz, A., Tseropoulos, G., Kerosuo, L., Mehrotra, P., Vk, B., ... Andreadis, S. T. (2019). Neural crest stem cells from human epidermis of aged donors maintain their multipotency in vitro and in vivo. *Scientific Reports, 9*, 9750. https://doi.org/10.1038/S41598-019-46140-9.

Mohlin, S., Kunttas, E., Persson, Cu, Abdel-Haq, R., Castillo, A., Murko, C., ... Kerosuo, L. (2019). Maintaining multipotent trunk neural crest stem cells as self-renewing crestospheres. *Developmental Biology*. https://doi.org/10.1016/J.Ydbio.2019.01.010.

Monsonego-Ornan, E., Kosonovsky, J., Bar, A., Roth, L., Fraggi-Rankis, V., Simsa, S., ... Sela-Donenfeld, D. (2012). Matrix metalloproteinase 9/gelatinase B is required for neural crest cell migration. *Developmental Biology, 364*(2), 162–177. https://doi.org/10.1016/J.Ydbio.2012.01.028).

Moore, R., Theveneau, E., Pozzi, S., Alexandre, P., Richardson, J., Merks, A., ... Mayor, R. (2013). Par3 controls neural crest migration by promoting microtubule catastrophe during contact inhibition of locomotion. *Development (Cambridge, England), 140*, 4763–4775.

Morin-Kensicki, E. M., & Eisen, J. S. (1997). Sclerotome development and peripheral nervous system segmentation in embryonic zebrafish. *Development (Cambridge, England), 124*(1), 159–167. https://doi.org/10.1242/Dev.124.1.159.

Morrison, S. J., Perez, S. E., Qiao, Z., Verdi, J. M., Hicks, C., Weinmaster, G., & Anderson, D. J. (2000). Transient notch activation initiates an irreversible switch from neurogenesis to gliogenesis by neural crest stem cells. *Cell, 101*, 499–510.

Motohashi, T., Watanabe, N., Nishioka, M., Nakatake, Y., Yulan, P., Mochizuki, H., ... Kunisada, T. (2016). Gene array analysis of neural crest cells identifies transcription factors necessary for direct conversion of embryonic fibroblasts into neural crest cells. *Biology Open, 5*, 311. https://doi.org/10.1242/Bio.015735.

Nakagawa, S., & Takeichi, M. (1995). Neural crest cell-cell adhesion controlled by sequential and subpopulation-specific expression of novel cadherins. *Development (Cambridge, England), 121*, 1321–1332.

Nakamura, H., & Ayer-Le Lievre, C. S. (1982). Mesectodermal capabilities of the trunk neural crest of birds. *Journal of Embryology and Experimental Morphology, 70*, 1–18.

Nardini, M., Ori, M., Vigetti, D., Gornati, R., Nardi, I., & Perris, R. (2004). Regulated gene expression of hyaluronan synthases during Xenopus laevis development. *Gene Expression Patterns: GEP, 4*(3), 303–308. https://doi.org/10.1016/J.Modgep.2003.10.006.

Nekooie Marnany, N., Fodil, R., Fereol, S., Dady, A., Depp, M., Relaix, F., ... Dufour, S. (2023). Glucose oxidation drives trunk neural crest cell development and fate. *Journal of Cell Science, 136*.

Newgreen, D. F. (1989). Physical influences on neural crest cell migration in avian embryos: Contact guidance and spatial restriction. *Developmental Biology, 131*(1), 136–148. https://doi.org/10.1016/S0012-1606(89)80045-4.

Newgreen, D. F., Ritterman, M., & Peters, E. A. (1979). Morphology and behaviour of neural crest cells of chick embryo in vitro. *Cell and Tissue Research, 203*, 115–140. https://doi.org/10.1007/Bf00234333.

Nieto, M. A. (2009). Epithelial-mesenchymal transitions in development and disease: Old views and new perspectives. *The International Journal of Developmental Biology, 53*, 1541–1547.

Nitzan, E., Pfaltzgraff, E. R., Labosky, P. A., & Kalcheim, C. (2013). Neural crest and schwann cell progenitor-derived melanocytes are two spatially segregated populations similarly regulated by Foxd3. *Proceedings of the National Academy of Sciences of the United States of America, 110*, 12709–12714.

Noden, D. M. (1975). An analysis of migratory behavior of avian cephalic neural crest cells. *Developmental Biology, 42*, 106–130.

Noden, D. M. (1983). The role of the neural crest in patterning of avian cranial skeletal, connective, and muscle tissues. *Developmental Biology, 96*, 144–165.

Ori, M., Nardini, M., Casini, P., Perris, R., & Nardi, I. (2006). Xhas2 activity is required during somitogenesis and precursor cell migration in Xenopus development. *Development (Cambridge, England), 133*(4), 631–640. https://doi.org/10.1242/Dev.02225.

Osborne, N. J., Begbie, J., Chilton, J. K., Schmidt, H., & Eickholt, B. J. (2005). Semaphorin/neuropilin signaling influences the positioning of migratory neural crest cells within the hindbrain region of the chick. *Developmental Dynamics: An Official Publication of the American Association of Anatomists, 232*, 939–949.

Paluch, E. K., Aspalter, I. M., & Sixt, M. (2016). Focal adhesion-independent cell migration. *Annual Review of Cell and Developmental Biology, 32*, 469–490.

Pei, D., Luther, W., Wang, W., Paw, B. H., Stewart, R. A., & George, R. E. (2013). Distinct Neuroblastoma-associated Alterations of PHOX2B Impair Sympathetic Neuronal Differentiation in Zebrafish Models. *PLoS Genetics, 9*(6), e1003533. https://doi.org/10.1371/journal.pgen.1003533.

Perissinotto, D., Iacopetti, P., Bellina, I., Doliana, R., Colombatti, A., Pettway, Z., ... Perris, R. (2000). Avian neural crest cell migration is diversely regulated by the two major hyaluronan-binding proteoglycans Pg-M/versican and aggrecan. *Development (Cambridge, England), 127*(13), 2823–2842. https://doi.org/10.1242/Dev.127.13.2823.

Perris, R. (1997). The extracellular matrix in neural crest-cell migration. *Trends in Neurosciences, 20*(1), 23–31. https://doi.org/10.1016/S0166-2236(96)10063-1.

Perris, R., & Perissinotto, D. (2000). Role of the extracellular matrix during neural crest cell migration. *Mechanisms of Development, 95*(1–2), 3–21. https://doi.org/10.1016/S0925-4773(00)00365-8.

Perris, R., Perissinotto, D., Pettway, Z., Bronner-Fraser, M., Mörgelin, M., & Kjmata, K. (1996). Inhibitory effects of PG-H/aggrecan and PG-M/versican on avian neural crest cell migration. *Federation of American Societies for Experimental Biology, 10*, 293–301. https://doi.org/10.1096/fasebj.10.2.8641562.

Poole, T. J., & Thiery, J. P. (1986). Antibodies and a synthetic peptide that block cell-fibronectin adhesion arrest neural crest cell migration in vivo. *Progress in Clinical and Biological Research, 217b*, 235–238.

Powell, D. R., Williams, J. S., Hernandez-Lagunas, L., Salcedo, E., O'brien, J. H., & Artinger, K. B. (2015). Cdon promotes neural crest migration by regulating N-cadherin localization. *Developmental Biology, 407*, 289–299.

Raible, D. W., & Eisen, J. S. (1994). Restriction of neural crest cell fate in the trunk of the embryonic zebrafish. *Development (Cambridge, England), 120*, 495–503.

Raible, D. W., & Eisen, J. S. (1996). Regulative interactions in zebrafish neural crest. *Development (Cambridge, England), 122*, 501–507.

Raible, D. W., Wood, A., Hodsdon, W., Henion, P. D., Weston, J. A., & Eisen, J. S. (1992). Segregation and early dispersal of neural crest cells in the embryonic zebrafish. *Developmental Dynamics: An Official Publication of the American Association of Anatomists, 195*, 29–42.

Reedy, M. V., Faraco, C. D., & Erickson, C. A. (1998). The delayed entry of thoracic neural crest cells into the dorsolateral path is a consequence of the late emigration of melanogenic neural crest cells from the neural tube. *Developmental Biology, 200*, 234–246.

Reedy, M. V., Johnson, R. L., & Erickson, C. A. (2003). The expression patterns of C-kit and Sl in chicken embryos suggest unexpected roles for these genes in somite and limb development. *Gene Expression Patterns: GEP, 3*, 53–58.

Replogle, M. R., Sreevidya, V. S., Lee, V. M., Laiosa, M. D., Svoboda, K. R., & Udvadia, A. J. (2018). Establishment of a murine culture system for modeling the temporal progression of cranial and trunk neural crest cell differentiation. *Disease Models & Mechanisms, 11*, Dmm035097. https://doi.org/10.1242/Dmm.035097.

Richardson, J., Gauert, A., Briones Montecinos, L., Fanlo, L., Alhashem, Z. M., Assar, R., ... Linker, C. (2016). Leader cells define directionality of trunk, but not cranial, neural crest cell migration. *Cell Reports, 15*, 2076–2088.

Ridley, A. J., Schwartz, M. A., Burridge, K., Firtel, R. A., Ginsberg, M. H., Borisy, G., ... Horwitz, A. R. (2003). Cell migration: Integrating signals from front to back. *Science (New York, N. Y.), 302*, 1704–1709.

Rios, A. C., Serralbo, O., Salgado, D., & Marcelle, C. (2011). Neural crest regulates myogenesis through the transient activation of notch. *Nature, 473*, 532–535.

Rocha, M., Singh, N., Ahsan, K., Beiriger, A., & Prince, V. E. (2020). Neural crest development: insights from the zebrafish. *Developmental Dynamics, 249*, 88–111. https://doi.org/10.1002/dvdy.122.

Rogers, C. D., Saxena, A., & Bronner, M. E. (2013). Sip1 mediates an E-cadherin-to-N-cadherin switch during cranial neural crest EMT. *The Journal of Cell Biology, 203*, 835–847.

Roth, L., Kalev-Altman, R., Monsonego-Ornan, E., & Sela-Donenfeld, D. (2017). A new role of the membrane-type matrix metalloproteinase 16 (Mmp16/Mt3-Mmp) in neural crest cell migration. *The International Journal of Developmental Biology, 61*(3–4–5), 245–256. https://doi.org/10.1387/Ijdb.160286ds.

Sadaghiani, B., & Thiebaud, C. H. (1987). Neural crest development in the Xenopus laevis embryo, studied by interspecific transplantation and scanning electron microscopy. *Developmental Biology, 124*, 91–110.

Santiago, A., & Erickson, C. A. (2002). Ephrin-B ligands play A dual role in the control of neural crest cell migration. *Development (Cambridge, England), 129*(15), 3621–3632. https://doi.org/10.1242/Dev.129.15.3621.

Sauka-Spengler, T., & Bronner-Fraser, M. (2008). A gene regulatory network orchestrates neural crest formation. *Nature Reviews. Molecular Cell Biology, 9*, 557–568.
Scarpa, E., Szabo, A., Bibonne, A., Theveneau, E., Parsons, M., & Mayor, R. (2015). Cadherin switch during EMT in neural crest cells leads to contact inhibition of locomotion via repolarization of forces. *Developmental Cell, 34*, 421–434.
Schilling, T. F., & Kimmel, C. B. (1994). Segment and cell type lineage restrictions during pharyngeal arch development in the zebrafish embryo. *Development (Cambridge, England), 120*, 483–494.
Schwarz, Q., Maden, C. H., Davidson, K., & Ruhrberg, C. (2009). Neuropilin-mediated neural crest cell guidance is essential to organise sensory neurons into segmented dorsal root ganglia. *Development (Cambridge, England), 136*(11), 1785–1789. https://doi.org/10.1242/Dev.034322.
Schweitzer, J., Becker, T., Lefebvre, J., Granato, M., Schachner, M., & Becker, C. G. (2005). Tenascin-C is involved in motor axon outgrowth in the trunk of developing zebrafish. *Developmental Dynamics: An Official Publication of the American Association of Anatomists, 234*(3), 550–566. https://doi.org/10.1002/Dvdy.20525.
Serbedzija, G. N., Bronner-Fraser, M., & Fraser, S. E. (1989). A vital dye analysis of the timing and pathways of avian trunk neural crest cell migration. *Development (Cambridge, England), 106*, 809–816.
Serbedzija, G. N., Bronner-Fraser, M., & Fraser, S. E. (1992). Vital dye analysis of cranial neural crest cell migration in the mouse embryo. *Development (Cambridge, England), 116*, 297–307.
Serbedzija, G. N., Burgan, S., Fraser, S. E., & Bronner-Fraser, M. (1991). Vital dye labelling demonstrates a sacral neural crest contribution to the enteric nervous system of chick and mouse embryos. *Development (Cambridge, England), 111*, 857–866.
Serbedzija, G. N., Fraser, S. E., & Bronner-Fraser, M. (1990). Pathways of trunk neural crest cell migration in the mouse embryo as revealed by vital dye labelling. *Development (Cambridge, England), 108*, 605–612.
Seto, Y., Ogihara, R., Takizawa, K., & Eiraku, M. (2024). In vitro induction of patterned branchial arch-like aggregate from human pluripotent stem cells. *Nature Communications, 15*, 1351. https://doi.org/10.1038/S41467-024-45285-0.
Shepherd, I., & Eisen, J. (2011). Development of the zebrafish enteric nervous system. *Methods in Cell Biology, 101*, 143–160.
Simoes-Costa, M., & Bronner, M. E. (2016). Reprogramming of avian neural crest axial identity and cell fate. *Science, 352*(6293), 1570–1573. https://doi.org/10.1126/science.aaf2729.
Snippert, H. J., & Clevers, H. (2011). Tracking adult stem cells. *EMBO Reports, 12*, 113–122.
Soldatov, R., Kaucka, M., Kastriti, M. E., Petersen, J., Chontorotzea, T., Englmaier, L., ... Adameyko, I. (2019). Spatiotemporal structure of cell fate decisions in murine neural crest. *Science (New York, N. Y.), 364*.
Stickney, H. L., Barresi, M. J., & Devoto, S. H. (2000). Somite development in zebrafish. *Developmental Dynamics: An Official Publication of the American Association of Anatomists, 219*, 287–303.
Su, Y., Naser, I. B., Islam, S. M., Zhang, S., Ahmed, G., Chen, S., ... Tanaka, H. (2009). Draxin, an axon guidance protein, affects chick trunk neural crest migration. *Development, Growth & Differentiation, 51*(9), 787–796. https://doi.org/10.1111/J.1440-169x.2009.01137.X.
Subkhankulova, T., Camargo Sosa, K., Uroshlev, L. A., Nikaido, M., Shriever, N., Kasianov, A. S., ... Kelsh, R. N. (2023). Zebrafish pigment cells develop directly from persistent highly multipotent progenitors. *Nature Communications, 14*, 1258.

Szabo, A., & Mayor, R. (2018). Mechanisms of neural crest migration. *Annual Review of Genetics, 52*, 43–63.

Szabó, A., Melchionda, M., Nastasi, G., Woods, M. L., Campo, S., Perris, R., & Mayor, R. (2016). In vivo confinement promotes collective migration of neural crest cells. *The Journal of Cell Biology, 213*(5), 543–555. https://doi.org/10.1083/Jcb.201602083.

Tan, S. S., & Morriss-Kay, G. M. (1986). Analysis of cranial neural crest cell migration and early fates in postimplantation rat chimaeras. *Journal of Embryology and Experimental Morphology, 98*, 21–58.

Taneyhill, L. A., Coles, E. G., & Bronner-Fraser, M. (2007). Snail2 directly represses Cadherin6b during epithelial-to-mesenchymal transitions of the neural crest. *Development (Cambridge, England), 134*, 1481–1490.

Testaz, S., Delannet, M., & Duband, J.-L. (1999). Adhesion and migration of avian neural crest cells on fibronectin require the cooperating activities of multiple integrins of the B1 and B3 families. *Journal of Cell Science, 112*, 4715–4728. https://doi.org/10.1242/Jcs.112.24.4715.

Theveneau, E., Marchant, L., Kuriyama, S., Gull, M., Moepps, B., Parsons, M., & Mayor, R. (2010). Collective chemotaxis requires contact-dependent cell polarity. *Developmental Cell, 19*, 39–53.

Theveneau, E., & Mayor, R. (2012). Neural crest delamination and migration: from epithelium-to-mesenchyme transition to collective cell migration. *Developmental Biology, 366*, 34–54.

Theveneau, E., Steventon, B., Scarpa, E., Garcia, S., Trepat, X., Streit, A., & Mayor, R. (2013). Chase-and-run between adjacent cell populations promotes directional collective migration. *Nature Cell Biology, 15*, 763–772.

Thisse, B., Heyer, V., Lux, A., Alunni, V., Degrave, A., Seiliez, I., ... Thisse, C. (2004). Spatial and temporal expression of the zebrafish genome by large-scale in situ hybridization screening. *Methods in Cell Biology,* 77, 505–519. https://doi.org/10.1016/S0091-679x(04)77027-2.

Thisse, B., Pflumio, S., Furthauer, M., Loppin, B., Heyer, V., Degrave, A., ... Thisse, C. (2001). *Expression of the Zebrafish Genome During Embryogenesis, Zfin Direct Data Submission.*

Tongiorgi, E., Bernhardt, R. R., Zinn, K., & Schachner, M. (1995). Tenascin-C mRNA is expressed in cranial neural crest cells, in some placodal derivatives, and in discrete domains of the embryonic zebrafish brain. *Journal of Neurobiology, 28*(3), 391–407. https://doi.org/10.1002/Neu.480280311.

Toole, B. P. (2001). Hyaluronan in morphogenesis. *Seminars in Cell & Developmental Biology, 12*(2), 79–87. https://doi.org/10.1006/Scdb.2000.0244.

Trainor, P. A. (2016). Developmental biology: We are all walking mutants. *Current Topics in Developmental Biology, 117*, 523–538.

Vega-Lopez, G. A., Cerrizuela, S., & Aybar, M. J. (2017). Trunk neural crest cells: Formation, migration and beyond. *The International Journal of Developmental Biology, 61*(1–2), 5–15. https://doi.org/10.1387/Ijdb.160408gv.

Wakamatsu, Y., Maynard, T. M., & Weston, J. A. (2000). Fate determination of neural crest cells by notch-mediated lateral inhibition and asymmetrical cell division during gangliogenesis. *Development (Cambridge, England), 127*, 2811–2821.

Wang, H. U., & Anderson, D. J. (1997). EPH family transmembrane ligands can mediate repulsive guidance of trunk neural crest migration and motor axon outgrowth. *Neuron, 18*(3), 383–396. https://doi.org/10.1016/S0896-6273(00)81240-4.

Wang, X., Chan, A. K., Sham, M. H., Burns, A. J., & Chan, W. Y. (2011). Analysis of the sacral neural crest cell contribution to the hindgut enteric nervous system in the mouse embryo. *Gastroenterology, 141*, 992–1002 E1-6.

Wehrle-Haller, B., Morrison-Graham, K., & Weston, J. A. (1996). Ectopic C-kit expression affects the fate of melanocyte precursors in patch mutant embryos. *Developmental Biology, 177*, 463–474.
Wehrle-Haller, B., & Weston, J. A. (1995). Soluble and cell-bound forms of steel factor activity play distinct roles in melanocyte precursor dispersal and survival on the lateral neural crest migration pathway. *Development (Cambridge, England), 121*, 731–742.
Wehrle-Haller, B., & Weston, J. A. (1997). Receptor tyrosine kinase-dependent neural crest migration in response to differentially localized growth factors. *Bioessays: News and Reviews in Molecular, Cellular and Developmental Biology, 19*, 337–345.
Wehrle-Haller, B., & Weston, J. A. (1999). Altered cell-surface targeting of stem cell factor causes loss of melanocyte precursors in steel17h mutant mice. *Developmental Biology, 210*, 71–86.
Weston, J. A., & Butler, S. L. (1966). Temporal factors affecting localization of neural crest cells in TBE chicken embryo. *Developmental Biology, 14*, 246–266.
Wiese, C. B., Deal, K. K., Ireland, S. J., Cantrell, V. A., & Southard-Smith, E. M. (2017). Migration pathways of sacral neural crest during development of lower urogenital tract innervation. *Developmental Biology, 429*, 356–369.
Williams, J. S., Hsu, J. Y., Rossi, C. C., & Artinger, K. B. (2018). Requirement of zebrafish Pcdh10a and Pcdh10b in melanocyte precursor migration. *Developmental Biology, 444*(1), S274–S286.
Wilson, H. C., & Milos, N. C. (1987). The effects of various nutritional supplements on the growth, migration and differentiation of Xenopus laevis neural crest cells in vitro. *In Vitro Cellular & Developmental Biology: Journal of the Tissue Culture Association, 23*, 323–331.
Wiszniak, S., & Schwarz, Q. (2019). Notch signalling defines dorsal root ganglia neuroglial fate choice during early neural crest cell migration. *BMC Neuroscience, 20*(1), 21. https://doi.org/10.1186/s12868-019-0501-0.
Workman, M. J., Mahe, M. M., Trisno, S., Poling, H. M, Watson, C. L., Sundaram, N., ... Wells, J. M. (2016). Engineered human pluripotent-stem-cell-derived intestinal tissues with a functional enteric nervous system. *Nature Medicine, 23*, 49. https://doi.org/10.1038/Nm.4233.
Yntema, C. L., & Hammond, W. S. (1945). Depletions and abnormalities in the cervical sympathetic system of the chick following extirpation of neural crest. *The Journal of Experimental Zoology, 100*, 237–263.
Yntema, C. L., & Hammond, W. S. (1954). The origin of intrinsic ganglia of trunk viscera from vagal neural crest in the chick embryo. *The Journal of Comparative Neurology, 101*, 515–541.
Zhang, S., Su, Y., Gao, J., Zhang, C., & Tanaka, H. (2017). A potential inhibitory function of draxin in regulating mouse trunk neural crest migration. *In Vitro Cellular & Developmental Biology. Animal, 53*(1), 43–53. https://doi.org/10.1007/S11626-016-0079-0.
Zimmerman, E., Geiger, B., & Addadi, L. (2002). Initial stages of cell-matrix adhesion can be mediated and modulated by cell-surface hyaluronan. *Biophysical Journal, 82*(4), 1848–1857. https://doi.org/10.1016/S0006-3495(02)75535-5 Pmid: 11916844; Pmcid: Pmc1301982.
Zirlinger, M., Lo, L., Mcmahon, J., Mcmahon, A. P., & Anderson, D. J. (2002). Transient expression of the BHLH factor neurogenin-2 marks a subpopulation of neural crest cells biased for a sensory but not a neuronal fate. *Proceedings of the National Academy of Sciences of the United States of America, 99*, 8084–8089.
Zuhdi, N., Ortega, B., Giovannone, D., Ra, H., Reyes, M., Asención, V., ... De Bellard Me (2015). Slit molecules prevent entrance of trunk neural crest cells in developing gut. *International Journal of Developmental Neuroscience: The Official Journal of the International Society for Developmental Neuroscience, 41*, 8–16. https://doi.org/10.1016/J.Ijdevneu.2014.12.003 Epub 2014 Dec 6. Pmid: 25490618; Pmcid: Pmc4355086.

CHAPTER FIVE

Cell migration during development: Is it always collective and never as a single cell?

Jorge Diaz, Amrutha Patkunarajah, and Roberto Mayor*
Department of Cell and Developmental Biology, Anatomy Building, University College London, London, United Kingdom
*Corresponding author. e-mail address: r.mayor@ucl.ac.uk

Contents

1. Introduction 154
2. Amoeboid and mesenchymal modes of cell migration 156
3. Single cell migration in development 159
   3.1 Primordial germ cell migration 159
   3.2 Zebrafish primordial germ cell migration 160
   3.3 *Drosophila melanogaster* primordial germ cell migration 163
4. Collective cell migration: epithelial and mesenchymal 166
5. Collective cell migration in development 168
   5.1 Lateral line in zebrafish 168
   5.2 Drosophila border cells 169
   5.3 Drosophila testis myotubes 173
6. Beyond collective cell migration: supracellular organization 174
   6.1 Neural crest migration in *Xenopus* 175
   6.2 Drosophila melanogaster gastrulation 176
7. Concluding remarks 177
Acknowledgments 179
References 179

Abstract

Cell migration is a fundamental process essential for homeostasis, disease progression, and developmental biology. This review explores the mechanisms of cell migration, focusing on both single-cell and collective migration modes, and highlights their roles in early development. We examine the characteristics of amoeboid and mesenchymal migration, emphasizing their regulation and implications across various classical developmental models. By analyzing examples of single-cell migration, such as primordial germ cell migration in zebrafish and *Drosophila melanogaster*, and examples of collective cell migration in the lateral line of zebrafish, border cells,

Current Topics in Developmental Biology, Volume 164
ISSN 0070-2153, https://doi.org/10.1016/bs.ctdb.2024.10.004

and testis myotubes in *Drosophila*, we illustrate the complexity and significance of cell-cell interactions, cell-matrix interactions, and the chemical and mechanical cues that drive migration. The review also highlights the "supracellular organization" observed in many systems where supracellular actomyosin cables are present, which allow for coordinated and cooperative movement. This cooperativity is crucial for effective positioning and function, ensuring proper tissue formation and responsive adaptation to environmental cues. This review provides insights and raises questions about the mechanisms of cell migration during development, supporting the idea that cells never migrate entirely alone. We propose that even in those cases normally described as single cell migration, some degree of collectiveness or cooperation is involved, suggesting that during development, cells always migrate in collective coordination when forming complex tissue and organ structures.

1. Introduction

The intricate process by which cells migrate from one location to another, with their distinct characteristics of directionality and persistence, has intrigued biologists for decades. Advances in microscopy, genetics, biophysics, and innovative techniques for visualizing cellular movements, such as live-cell imaging and cell tracking technologies, have intensified interest in studying this fundamental biological phenomenon. Cell migration plays a crucial role in maintaining homeostasis, facilitating organismal development, and playing a fundamental role in disease progression (Friedl & Gilmour, 2009; Rørth, 2011). Gaining a deeper understanding of how cell migration functions is crucial for immunology and cancer research. During development, cell migration is essential to understand how different tissues and organs are shaped during morphogenesis.

During embryonic development, cell migration is essential for forming and organizing complex tissues and organs. For example, during gastrulation, cells migrate to form the body's three primary tissue layers: ectoderm, mesoderm, and endoderm, which are foundational for developing all body tissues and organs (Gerhart & Keller, 1986). Disruptions in cell migration during development can lead to significant problems, such as neural tube defects and facial abnormalities (Alvizi et al., 2023), among many others. Neural crest cells are particularly remarkable in this context because they undergo epithelial-mesenchymal transition (EMT) and migrate extensively throughout the embryo in a collective manner. They finally differentiate into various cell types, including those of the peripheral nervous system, craniofacial cartilage and bone, teeth, pigment cells (melanocytes), parts of the cardiovascular system, adrenal medulla cells, connective tissue, and endocrine cells. Accordingly, neural crest cells serve as a valuable model for studying cell migration in development due to their extensive and varied migration patterns, making them a focal point of intense and innovative research (Mayor & Theveneau, 2013).

In maintaining homeostasis, cell migration is essential for tissue repair, renewal of cellular structures, and responding to physiological needs. For instance, epithelial cells in the skin, gut, and respiratory tract migrate continuously to restore and maintain the epithelial barrier, which is crucial for defending against pathogens and preserving tissue integrity (Rørth, 2009). The intestinal epithelium provides a particularly well-known example of epithelial migration. Cells migrate from the crypts to the villi to sustain gut homeostasis. Studies in vivo and in intestinal organoids have shown that cells in the crypt generate mechanical tension that controls their migration, with forces distributed along the epithelial structure shaping the crypt-villus architecture (Pérez-González et al., 2021). Additionally, endothelial cells of the blood-brain barrier migrate to maintain selective permeability, regulating the exchange of molecules between the bloodstream and brain tissue, which is essential for correct brain function (Engelhardt & Liebner, 2014).

Pathologically, abnormal cell migration plays a significant role in various diseases. For example, when vascular smooth muscle cells migrate improperly, it can lead to atherosclerosis and changes in blood vessels associated with cardiovascular disorders (Gimbrone & García-Cardeña, 2016). Similarly, in autoimmune diseases such as rheumatoid arthritis, unregulated migration of immune cells leads to chronic inflammation and tissue damage (Alivernini, Firestein, & McInnes, 2022). Inflammatory signals direct immune cells to sites of infection or injury, which is essential for an effective immune response but can be harmful if not properly regulated (Mantovani, Allavena, Sica, & Balkwill, 2008). These signals also cause astrocytes, the most abundant cells in the brain, to become reactive and migrate to injury sites, playing a critical role in tissue repair and neuronal protection (Sofroniew, 2015). In cancer, cell migration enables tumor cells to detach from the primary tumor, invade the extracellular matrix (ECM), and establish secondary tumors at distant sites (Hanahan, 2022; Joyce & Pollard, 2009).

Cells can migrate using different modes depending on if they migrate as single cells or a group of cells, their mode of migration can also vary according to interaction with the substrate and the mode of locomotion. In this review, we will focus on amoeboid and mesenchymal migration and single and collective migration.

2. Amoeboid and mesenchymal modes of cell migration

When cells migrate, they utilize different modes to sense their environments and optimize their movement. These strategies involve dynamic changes in cell shape and the reorganization of both cell-cell and cell-matrix interactions. Fundamentally, these modes are governed by the adhesive properties of surfaces and the contractile capabilities of the cells. Amoeboid and mesenchymal migration modes represent distinct mechanisms shaped by these cellular interactions with their surroundings.

Amoeboid cell migration (Fig. 1A) is characterized by cells adopting a rounded and deformable shape, often accompanied by the formation of blebs. Blebs are balloon-like cell membrane protrusions caused by localized weakening or detachment of the cell cortex or its connections to the membrane, which inflate and deflate through internal pressure or water influx (García-Arcos, Jha, Waterman, & Piel, 2024). This unique feature promotes fast movements driven by actomyosin contractility, generating hydrostatic pressure to push the cell forward (Charras & Paluch, 2008; Lämmermann & Sixt, 2009). As a result, cells with amoeboid features migrate at 2–25 μm/min in 3D environments, as observed in A375m2 melanoma cells and lymphocytes in collagen gel. This is significantly faster than mesenchymal migration, which occurs at approximately 0.1–0.5 μm/min in 3D matrices (Paňková, Rösel, Novotný, & Brábek, 2010).

An important aspect regarding amoeboid migration is confinement, which refers to the physical restriction of cell movement within narrow spaces in tissues. In these confined environments, amoeboid cells can more easily squeeze and deform themselves to pass through narrow gaps, with minimal remodeling of the extracellular matrix (ECM) (García-Arcos et al., 2024). Confinement has been shown to influence amoeboid migration by promoting the formation of blebs. This is supported by the evidence from cancer cells migrating in 3D matrices, where bleb formation is particularly pronounced under confinement (Sahai & Marshall, 2003; Wolf et al., 2003). Similar phenomena were observed in zebrafish primordial germ cells, where cells migrate through narrow spaces, highlighting the role of confinement in enhancing amoeboid migration, as we will discuss later (Liu et al., 2015; Ruprecht et al., 2015; Schick & Raz, 2022). Amoeboid migration is commonly observed in immune cells, such as leukocytes, as they navigate through tissues to reach sites of infection or injury.

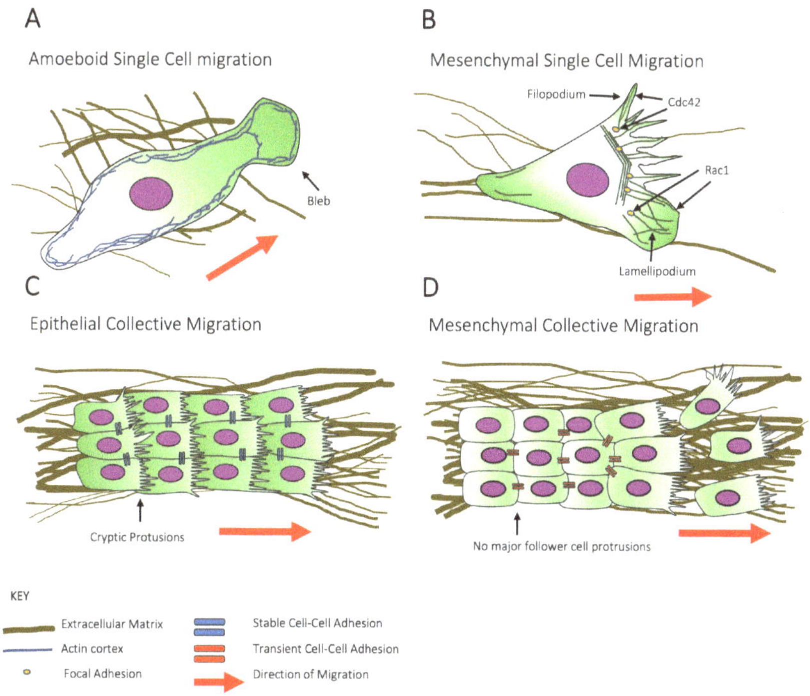

Fig. 1 From single cell to collective migration in amoeboid and mesenchymal modes. (A) Amoeboid single cell migration shows a rounded morphology, crucially driven by actomyosin contractility that generates internal pressure, facilitating rapid movement through the ECM without extensive remodeling. This mode of migration often involves the formation of blebs, which are actin cortex detachments that expand due to intracellular pressure and contribute to rapid movement. (B) Mesenchymal cells elongate and adhere strongly to the ECM using lamellipodia and filopodia to remodel the matrix. Rac1 regulates the formation of lamellipodia, while Cdc42 promotes filopodia extension, both of which remodel the ECM and guide the direction of migration. (C) Epithelial collective migration moves as a cohesive sheet, maintaining stable cell-cell adhesions and polarity. Cryptic protrusions that form between cells play a crucial role in environmental sensing, guiding, and coordinating the direction of the migration. (D) Mesenchymal collective migration involves the loss of cell polarity and cells move as a cluster with transient cell-cell adhesions. These temporary cell-cell connections can influence the formation of protrusion at the leading edge and contribute to the overall directional movement. Protease activity plays a critical role in remodeling the ECM as the cell cluster moves.

Additionally, dendritic cells, key players in the immune system, utilize amoeboid migration to move from peripheral tissues to lymph nodes, where they present antigens to T cells, initiating adaptive immune responses (Randolph, Angeli, & Swartz, 2005).

In contrast, mesenchymal migration (Fig. 1B), one of the most studied modes of cell movement, involves cells adopting an elongated shape characterized by prominent protrusions like lamellipodia and filopodia governed by Rac1 and Cdc42, respectively, as well as integrin signaling at their leading edge (Nobes & Hall, 1995; Ridley, Paterson, Johnston, Diekmann, & Hall, 1992). Rac1 and Cdc42 activation happens in response to chemoattractant signals and integrin engagement with ECM, which incorporate external environmental signal into actin polymerization and cytoskeletal remodeling. These protrusions are driven by actin polymerization and are essential for cell attachment to the ECM and contribute to the generation of traction forces. Mesenchymal migration relies on strong cell-ECM interactions. Cells adhere to the ECM via integrin receptors, generating traction forces that pull the cell body forward. Protease activity remodels the ECM, allowing cells to create paths through the tissue (Hynes, 2009). This mode is slower than amoeboid migration but highly directional and involves significant ECM remodeling and strong adhesive interactions. Examples of cells that migrate with the mesenchymal program include fibroblasts, smooth muscle cells, chondrocytes, osteoblasts, pericytes, and adipocytes, each contributing to specific tissue functions (Kalluri & Weinberg, 2009). In cancer, mesenchymal cells often undergo phenotypic changes known as mesenchymal transition, contributing to tumor progression and metastasis. These cells acquire enhanced migratory and invasive capabilities, facilitating their dissemination from the primary tumor site to distant locations in the body (Hanahan, 2022).

These two modes of migration -mesenchymal and amoeboid- depend on the interaction with the substrate: the mesenchymal mode of migration involves focal adhesions and traction forces, where cells adhere to the surrounding ECM and apply pulling forces. In contrast, the amoeboid migration is characterized by adhesion-independent contractility. The prevalence of these two modes in a particular cell is determined by the capacity of the cell to adhere to the ECM and their degree of contractility. These mechanisms play a crucial role during development, allowing cells to migrate either in a mesenchymal or amoeboid manner, either individually or collectively, depending on the specific biological context. In the following sections of this review, we will provide examples of mesenchymal or amoeboid migration in both single and collective migration in classical development models such as *Drosophila*, zebrafish, and *Xenopus*.

3. Single cell migration in development

During development, cells predominantly migrate as a collective, but few single cell examples have been described in vivo. The clearest examples of single cell migration in development correspond to the movement of germ cells in *Drosophila* and zebrafish embryos, which we will discuss here (Aman & Piotrowski, 2010). In these models, germ cells undergo a "trans-embryonic" migration, meaning they originate far from their final destination in the gonads and must travel long distances across the embryo to reach their target. It is important to highlight that these cells use an amoeboid mode of migration, which might suggest that amoeboid migration occurs only as a single cell. In this context, an amoeboid mode of migration is favorable for cells to travel such long distances at an increased speed, ensuring that cell differentiation occurs at the correct target locations of the embryo without disrupting tissue function (Paluch & Raz, 2013).

3.1 Primordial germ cell migration

Primordial germ cells are a type of stem cells that are precursors to the gametes (oocytes and spermatozoa). The migration of primordial germ is a well-studied phenomenon that occurs in early development. During this process, the germ cells must migrate to the target site to complete the process of differentiation into gametes (Richardson & Lehmann, 2010). In this case, the target site would be the location where the prospective gonadal organs arise. Most of our understanding of primordial germ cell migration has been established by studies in zebrafish, *Drosophila melanogaster* and mice. This is mainly because of the markers that allow the identification and tracking of germ cells during migration (Barton, Roa-De La Cruz, Lehmann, & Lin, 2024). Although there are differences in the molecular mechanisms underlying primordial germ cell migration between organisms, the steps in the migration process are fundamentally similar across all organisms. In the first stage, the germplasm that contains the determinants for germ cell differentiation becomes specified as primordial germ cells by cell-autonomous mechanisms or by signaling via cell-cell contacts that instruct them to become germ cells, instead of somatic cells. At this stage, the cells are immotile (Richardson & Lehmann, 2010). Primordial germ cells become motile due to increased expression and activation of chemokine receptors in response to some developmental signals. These cells can now respond to directional cues during migration. The presence of chemo-attractants and chemo-repellents determines the migratory paths followed by these cells

(Raz, 2004; Richardson & Lehmann, 2010). This section discusses evidence in the literature and raises the question of whether primordial germ cells migrate truly as single cells.

3.2 Zebrafish primordial germ cell migration

In zebrafish, the specification of primordial germ cells begins with the formation of germplasm at the approximately 4-cell stage of zebrafish embryo development. Maternally produced mRNAs, such as *vas* and *nanos-1*, are expressed during this process (Knaut, Pelegri, Bohmann, Schwarz, & Nüsslein-Volhard, 2000; Köprunner, Thisse, Thisse, & Raz, 2001; Yoon, Kawakami, & Hopkins, 1997). At approximately 1000-cell stage, four random cell clusters start to form at four different regions of the embryo. Once the primordial germ cells are specified, these cells acquire motility and are prepared to start migrating to their target location. The migration path is quite complex and is comprised of multiple intermediate target locations before eventually reaching the final gonadal region for differentiation. It is known that migration is guided by the presence of stromal cell-derived factor 1A (SDF-1A), which is expressed at the intermediate and target sites of the embryo. First, the primordial germ cells migrate to the dorsal side of the developing embryo, which occurs during gastrulation at the shield stage (Fig. 2A) (Weidinger, Wolke, Köprunner, Klinger, & Raz, 1999). During segmentation (1–4 somites stage), primordial germ cells migrate to intermediate locations (Fig. 2B), then migrate to their final location at 5–9 somites (Fig. 2C), where gonads are formed (Fig. 2D) (Weidinger et al., 1999). When the primordial germ cells reach the final target tissue to form the gonads, cells become immotile, and migration ceases at a location that exhibits a high level of SDF-1A (Reichman-Fried, Minina, & Raz, 2004). The consensus regarding primordial germ cell migration in zebrafish is that migration occurs as single cell.

One of the key aspects of primordial germ cell migration is cell morphology and the mode of migration. A mechanistic study determined the factors that caused immotile primordial germ cells to become motile: (1) Cells formed blebs when the RNA-binding protein Dead-end (Dnd), was expressed, increasing myosin-dependent cell contractility, (2) E-cadherin expression was regulated by Dnd expression, which in turn regulated the expression of zeb1, a transcription factor, this leads to reduced cell-cell adhesion, and (3) the reduction of membrane-cell cortex attachment via the inhibition of Annexin A5b activity, which is a scaffold calcium- and phospholipid-binding protein (Goudarzi et al., 2012). Furthermore, bleb

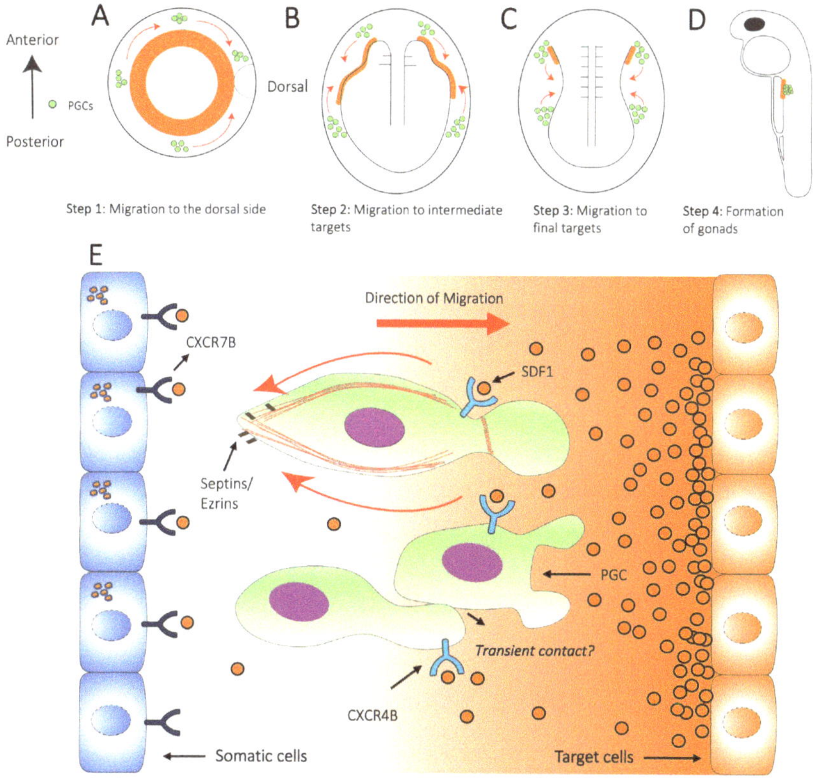

Fig. 2 The stages of primordial germ cell (PGC) migration in zebrafish, which occurs after the specification of the germ cells. The gradient of chemoattractant SDF-1A is indicated in orange. (A) highlights the migration of PGCs to the dorsal side of the embryo at the Shield stage of zebrafish development. (B) indicates that at the 2-somite stage, PGCs migrate to intermediate locations guided by chemotaxis towards SDF-1a. At 1–4 somite stage, (C), the PGCs migrate to their final location, where gonads form after 24 hpf (D). (E) Mechanisms underlying the migration of zebrafish PGCs. Migration is directed by the gradient of the chemoattractant SDF-1A. Cells are known to undertake an amoeboid mode of migration, as highlighted by bleb formation via Rac1/RhoA signaling at the leading edge of the cell. There is a retrograde flow of actin within the migrating cell and at the rear of the cell, bleb formation is inhibited by septins and ezrins. Somatic cells express the receptor, CXCR7b, which is reported to internalize and degrade SDF-1a to enhance the chemotactic gradient.

formation along the direction of migration was due to myosin-mediated contraction (Blaser et al., 2006). In addition to the bleb inflation and actin polymerization, the actomyosin contractility at the rear part of the cell drives the retrograde flow of cortical actin, which moves proteins such as ezrin (a protein that connects the actin cortex to the plasma membrane) to the back of the cell, where bleb formation is inhibited (Fig. 2E) (Barton et al., 2024; Olguin-Olguin et al., 2021). Bleb formation is a characteristic of amoeboid migration, and this mode is often associated with single-cell migration. Therefore, these findings suggested a more single cell behavior during primordial germ cell migration. Since primordial germ cells are initially organized in clusters, questions arise about these cells establishing transient cell-cell contacts during migration. It has been shown the downregulation of membrane E-cadherin was important for primordial germ cell motility, as E-cadherin mediates adhesion between germ and somatic cells generating the traction forces required for their migration (Blaser et al., 2005; Kardash et al., 2010). Therefore, these observations show that primordial germ cells establish cell-cell contacts with the somatic cells, but they do not allow us to completely rule out that transient contacts are established between the E-cadherin-expressing germ cells.

Another important characteristic of zebrafish primordial germ cell migration is directional migration towards the chemoattractant SDF-1A, which acts as a ligand for the CXCR4b receptor (Doitsidou et al., 2002). As the primordial germ cells migrate, SDF-1A is expressed in somatic tissues, becoming a cue for migrating primordial germ cells to reach their intermediate targets (Blaser et al., 2006). SDF-1A is also expressed in the target location, which is the gonadal region of the embryo (Blaser et al., 2005). However, the lack of the receptor CXCR4b or SDF-1A does not disrupt the migration of primordial germ cells, rather the lack of either leads to random migration within the embryo (Reichman-Fried et al., 2004). These observations emphasize the importance of SDF-1A and the receptor CXCR4b in guiding the migration of primordial germ cells to their target locations. Evidence from other systems suggests that SDF-1A work as a chemoattractant during germ cell migration, and if so, cells should follow a gradient of SDF-1A. The SDF-1A gradient is generated by the activity of CXCR7 which works as a decoy receptor, as it binds to and internalizes SDF-1A by sequestering the chemoattractant inside the cell and subsequently clearing the SDF-1A from the microenvironment, a mechanism shown in many contexts within zebrafish (Boldajipour et al., 2008; Bussmann & Raz, 2015; Cubedo, Cerdan, Sapede, & Rossel, 2009; Donà et al., 2013; Mahabaleshwar, Tarbashevich, Nowak, Brand, & Raz, 2012; Venkiteswaran et al., 2013) (Fig. 2E). CXCR7 is expressed in somatic cells where it acts as a sink for SDF-1A, allowing the generation of SDF-1 gradient required to direct germ cell migration (Boldajipour et al., 2008; Mahabaleshwar et al., 2012).

3.3 *Drosophila melanogaster* primordial germ cell migration

Similar to primordial germ cell migration in zebrafish, *Drosophila melanogaster* also undergoes a process that includes primordial germ cell specification, initiation of motility, and the establishment of migratory paths to the target location. *Drosophila* primordial germ cells are formed by undergoing cellularization of the germplasm at the posterior pole during early development (Santos & Lehmann, 2004). The germplasm consists of granule mRNAs such as *nanos* (*nos*), *polar granule component* (*pgc*) and *germ cell-less* (*gcl*) that are vital for the specification of germ cells (Santos & Lehmann, 2004; Trcek & Lehmann, 2019). In *Drosophila* embryos at stages 4 to 5 (i.e. 1.5 to 3 h after egg laying (hAEL)), approximately 35 primordial germs cells arise on the posterior side of the embryo, neighboring the developing somatic cells of the posterior midgut primordium (Jaglarz & Howard, 1994; Richardson & Lehmann, 2010). Motility initiation commences at approximately 2 to 2.5 hAEL, where primordial germ cells become migratory. Similar to zebrafish primordial germ cells, *Drosophila* primordial germ cells form cell clusters within the lumen of the posterior midgut (Jaglarz & Howard, 1995; Kunwar et al., 2008). The initiation of *Drosophila* primordial germ cell migration is modulated by signaling via Trapped in endoderm (TRE1), a G protein-coupled receptor. This leads to cell polarization, resulting in a radial organization where the leading edge of the cells faces outwards towards the midgut primordium, while the tails of the cells are at the core of the cluster (Kunwar et al., 2008). Finally, cells become fully motile when they lose their intercellular adhesions, and protrusions extend towards epithelial cells of the midgut primordium (Richardson & Lehmann, 2010). At stage 10, the newly motile germ cells extravasate out of the epithelial layer of the midgut (Fig. 3A), a process regulated by the G protein-coupled receptor TRE1, which activates Rho signaling, causing changes in the cytoskeletal arrangement (Fig. 3E). Then at stage 11, the cells migrate along the surface of the midgut (Fig. 3B). The cells then enter the mesoderm and undergo bilateral migration towards the somatic gonadal precursors (SGPs) at stage 12 (Fig. 3C), and after stage 13, the gonads form (Fig. 3D) (Jaglarz & Howard, 1995; Kunwar et al., 2008). The expression of both transmembrane lipid phosphate phosphatases, Wunen (WUN) and Wunen-2 (WUN-2), and 3-hydroxy-3-methylglutaryl coenzyme A reductase (HMGCR), and the presence of phospholipids guide germ cell migration into the mesoderm (Santos & Lehmann, 2004). A loss of WUN and WUN-2 can result in a drastic disruption in migration

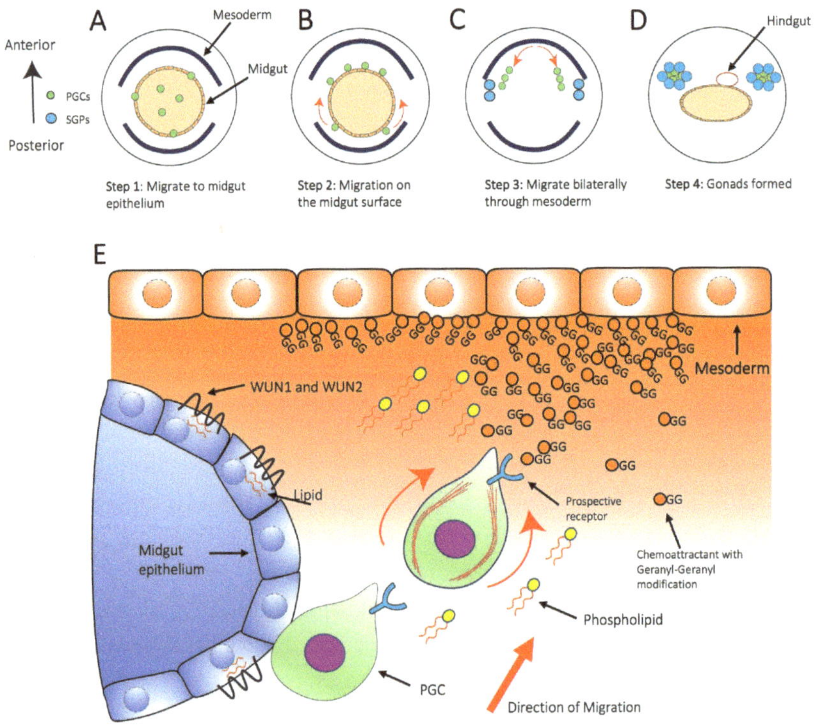

Fig. 3 The stages of PGC migration in *Drosophila melanogaster*, which are depicted by diagrams of transverse sections of the developing embryo. At stage 10, PGCs migrate through the midgut epithelium (A). This is followed by the migration of PGCs on the midgut surface between stages 11 and 12 (B), after which they migrate bilaterally through the mesoderm at stages 12 and 13 (C). Finally, the gonads are formed after stage 13 (D). (E) Similar to zebrafish PGCs, there is a retrograde flow of actin in *Drosophila* PGCs. However, there is no reported bleb formation. The expression of WUN1 and WUN2 in regions that repel PGCs, such as the midgut epithelial cells, regulates the directed migration of PGCs toward the mesoderm. WUN1 and WUN2 hydrolyze an extracellular phospholipid that serves as a PGC attractant and survival factor. PGCs are attracted to the mesoderm, where somatic gonad precursors (SGPs) release a chemoattractant modified by geranyl-geranyl (GG). At the rear of the migrating PGCs, Rho signaling and actomyosin contractility contribute to their movement.

(Sano, Renault, & Lehmann, 2005; Zhang, Zhang, Purcell, Cheng, & Howard, 1997). However, an overexpression of WUN and WUN-2 acts as a repellent in primordial germ cell migration (Zhang et al., 1997). The HMGCR pathway is responsible for attaching a geranyl-geranyl (GG) group to the chemoattractant at the SGPs and the secretion of the chemoattractant depends on the expression of ATP-binding cassette (ABC) transporter, multidrug resistance 49 (MDR49) (Ricardo & Lehmann, 2009). A recent study showed that the chemoattractant signal downstream of HMGCR is long-range, which can attract germ cells at least a distance of approximately 51 μm (Kenwrick, Mukherjee, & Renault, 2019). The same study reported that the HMGCR-mediated attraction can simultaneously function with the chemo-repellent pathway mediated by WUN/WUN-2, however, the latter was not a requirement for the function of the chemo-attraction pathway (Kenwrick et al., 2019). The overall above evidence suggests that migration is guided by the expression of chemoattractants and repellents.

In contrast to zebrafish, *Drosophila* primordial germ cells exhibit a round, spherical cell morphology when migrating in vivo. However, similar to zebrafish primordial germ cells, actin polymerizes in the direction of migration and actomyosin contractility is present at the rear part of the cell (Lin, Luo, & Lehmann, 2022). This means that the actomyosin contractility at the back of the cell causes a local pulling force on the cortical actin network, which triggers retrograde cortical actin flow. Undirected migration of *Drosophila* primordial germ cells usually occurs when cells are beginning to migrate towards the mesoderm (after endoderm egression) and when cells are very close to the SGPs (Lin et al., 2022). At these times, cells usually spin into place, and this is due to the cortical flows of actin being directed to the cell periphery in a circular motion. Likewise, in vitro experiments of isolated *Drosophila* primordial germ cells showed that on glass, 41 % of cells exhibited circular flows of actin and myosin (Lin et al., 2022), suggesting that the spherical morphology of *Drosophila* primordial germ cells and amoeboid mode of migration is not a response to the environment properties. Despite this observation, there is increased RhoA activity at the cell periphery but there is no polarization of Rac1 activity (Lin, Luo, & Lehmann, 2020). Whilst these two examples of primordial germ cell migration exhibit similar cell behavior, such as an amoeboid mode of migration, they utilize different signaling pathways that regulate the external cues that guide the directionality of migration.

In both zebrafish and *Drosophila* germ cells, the evidence suggests that they migrate as individual cells, however, the alternative hypothesis cannot be completely ruled out; for example, germ cells express cell-cell adhesion molecules on their surface which will allow some degree of transient interaction between them, as it has been shown for the mice germ cells (Barton et al., 2024; Massimo De, Scaldaferri, & Farini, 2005; Richardson & Lehmann, 2010), leading to a form of collective migration. Additionally, both kinds of germ cells respond to chemoattractants and it is likely that when the chemotactic molecules bind to the cell receptor it will become internalized, leading to a reduction in the level of the chemoattractant in the extracellular environment. This modification in the level of the chemoattractant could affect the shape of a gradient which will be sensed by neighbor cells affecting their responses and fulfilling one of the definitions of collective cell migration, which is cooperation among cells. Further experimental analysis is required to test whether germ cells exhibit some degree of collective migration.

4. Collective cell migration: epithelial and mesenchymal

Biologists often prioritize studying the migration of single cells due to the precision and depth of information provided by advanced techniques such as single-cell tracking and genetic manipulation. However, focusing exclusively on single-cell migration may introduce a bias, as cells rarely migrate in isolation in natural environments, especially during embryo development. Instead, they typically move collectively or in cooperation with other cells and tissues. By studying collective cell migration, biologists can better understand the complex interactions and signaling pathways that drive these coordinated movements, leading to a more comprehensive understanding of cellular dynamics. Developmental biology models are particularly interesting for studying collective cell migration because they integrate complex cell movements and tissue interactions to generate tissues and organs.

Collective migration could be defined as a fundamental process in biology where groups of cells move cohesively in a cooperative manner, maintaining both physical and functional connections with neighboring cells. The type of collective cell migration is influenced by a combination that depends on how strong cell-cell interactions and environmental cues are. Adherens junctions (AJs) and tight junctions (TJs) are crucial for coordinating collective migration, ensuring synchronized movement while preserving tissue integrity (Friedl & Gilmour, 2009). Environmental cues, such as chemical gradients and mechanical signals from the extracellular matrix, guide migration direction and speed, facilitating coordinated movement towards specific targets in response to developmental cues and wound healing signals (Hynes, 2009; Ridley et al., 2003).

The formation of leader and follower cells is another characteristic of collective cell migration architecture. Cells in the collective integrate environmental cues and adjust their cell-cell interactions to promote adhesion and generate forces essential for cohesive cluster migration. Leader cells often interact with their environment at the migration forefront, forming protrusions that promote focal adhesions and force generation to propel clusters forward, but their specific roles can vary depending on the context (Campanale & Montell, 2023). Concurrently, they establish cadherin-based AJs with neighboring cells, critical for cluster cohesion. This coupling of cell-cell and cell-matrix interactions not only involves mechanical aspects but also contributes to polarization within leader cells themselves, though the extent of their role in force generation differs across systems. Front-rear polarization leads to spatial organization of the cytoskeleton and cytoplasm, enabling directional protrusions at the leading edge while retracting the rear of the cell. While these mechanisms generally hold true, diverse models and recent research underscore variations in the roles of leader and follower cells in terms of force generation, cytoskeletal rearrangement, polarization, and molecular localization (Shellard & Mayor, 2019). As highlighted by Campanale and Montell (Campanale & Montell, 2023), the relative contributions of leaders and followers differ depending on the system, and follower cells can be equally, if not more, essential for driving collective migration forward. For example, in some contexts, followers play an active role in pushing leader cells forward, rather than simply being pulled. This has been observed in the zebrafish posterior lateral line primordium, where contraction at the rear of the primordium applies higher traction force than that exerted by the leader cells alone (Yamaguchi et al., 2022). Similarly, in some epithelial sheets, follower cells actively switch for leader positions, generating a dynamically switching roles to sustain collective migration (Campanale & Montell, 2023).

To explore the diverse mechanisms of collective cell migration, we will categorize them into epithelial and mesenchymal collective cell migration. Epithelial collective cell migration (Fig. 1C) predominantly occurs in epithelial sheets, where cells maintain junctions and apicobasal polarity, moving cohesively. This migration involves maintaining epithelial polarity and AJs intact, controlled by cadherin-dependent intercellular adhesion. Cryptic protrusions, subtle extensions between epithelial cells, play a significant role by sensing the environment and facilitating intercellular communication (Farooqui & Fenteany, 2005; Guillot & Lecuit, 2013).

In some cases, epithelial cells undergo a process known as EMT, during which cells gradually lose their epithelial characteristics, such as apicobasal polarity and strong cell-cell adhesion, and gain increased motility. Surprisingly, even after EMT cells can maintain transient interactions and migrate collectively, in what is known as mesenchymal collective cell migration (Fig. 1D). In mesenchymal collective migration cells adopt a mesenchymal phenotype and migrate in clusters through the extracellular matrix using protease-mediated degradation and traction forces. During EMT, cells downregulate epithelial markers like E-cadherin and upregulate mesenchymal markers such as N-cadherin and vimentin, which facilitate their interaction with the matrix and other cells (Kalluri & Weinberg, 2010; Thiery, Acloque, Huang, & Nieto, 2009; Wolf & Friedl, 2006). Integrins, which bind to ECM proteins, play a crucial role in mesenchymal migration by mediating adhesion and transmitting mechanical signals that regulate cell movement (Friedl & Wolf, 2010; Hynes, 2002).

Both epithelial and mesenchymal collective cell migration involve coordinated multicellular interactions with the microenvironment. Notable examples of collective migration are those observed during embryonic development, where cells migrate cooperatively to form complex tissues and organs.

5. Collective cell migration in development

Examples of collective cell migration during development include the lateral line system in *zebrafish*, border cells, and testis myotubes in *Drosophila*, among many others (for reviews on collective cell migration see: Friedl & Gilmour, 2009; Mishra, Campanale, Mondo, & Montell, 2019; Saraiva & Barriga, 2021; Scarpa & Mayor, 2016; Schumacher, 2019). These collective migrations involve intricate interactions that include cell-cell junctions and extracellular matrix components, ensuring that cells reach their correct destinations and contribute to tissue integrity and function. This section of the review will focus on these examples, highlighting the mechanisms and significance of collective cell migration in developmental contexts.

5.1 Lateral line in zebrafish

In the lateral line of zebrafish, cells migrate in a coordinated manner to form a sensory system along the body axis, which is crucial for their proper development and function. The development of the lateral line organ in zebrafish is a prime example of collective cell migration during embryogenesis (Fig. 4A and B). This process closely resembles epithelial collective cell migration because the migrating cells form a cohesive group that maintains strong cell-cell junctions. However, the leader cells adopt a mesenchymal-like behavior, characterized by front-rear polarity and prominent lamellipodial and filopodial protrusions. The zebrafish lateral line system is especially valuable for research as amenable to in vivo imaging and genetic manipulability. It comprises mechanosensory hair cell organs known as neuromasts, which are distributed across the skin by the posterior lateral line primordium (pLLP), a unified group of more than 100 migrating cells. Originating from neurogenic placodes near the otic vesicle, the pLLP undergoes coordinated migration along specific routes, depositing clusters of neuromasts and organizing them into interconnected rosettes. These structures collectively form the functional lateral line organ (Ghysen & Dambly-Chaudière, 2004; Haas & Gilmour, 2006).

The path followed by the pLLP is guided by chemokine gradients, particularly the zebrafish homolog of stromal-derived factor 1 (SDF1). SDF1 is expressed uniformly with no apparent gradient, instead, the gradient is self-generated by the migrating primordium. The primordium expresses two kinds of SDF1 receptors: CXCR4 is expressed in the anterior cells where it activates a signal that leads to forward migration, whereas CXCR7 is expressed in the posterior cells where it binds and internalizes SDF1, but without signaling activation. CXCR4 works as a decoy receptor that clears SDF1 from the environment generating an anterior-posterior gradient around the migrating primordium (Lau et al., 2020; Valentin, Haas, & Gilmour, 2007). Once positioned, the neuromasts undergo further maturation into functional sensory organs capable of detecting mechanical disturbances in the surrounding water environment. Each neuromast comprises mechanosensory hair cells and support cells derived from progenitors within the migrating pLLP. The differentiation process involves dynamic changes in gene expression profiles, transitioning from multipotent progenitors to committed sensory lineages under the influence of local signaling cues and extracellular matrix interactions (Ghysen & Dambly-Chaudière, 2004; Haas & Gilmour, 2006).

5.2 Drosophila border cells

In *Drosophila melanogaster*, the migration of border cells within the egg chamber during oogenesis is a clear example of collective cell migration

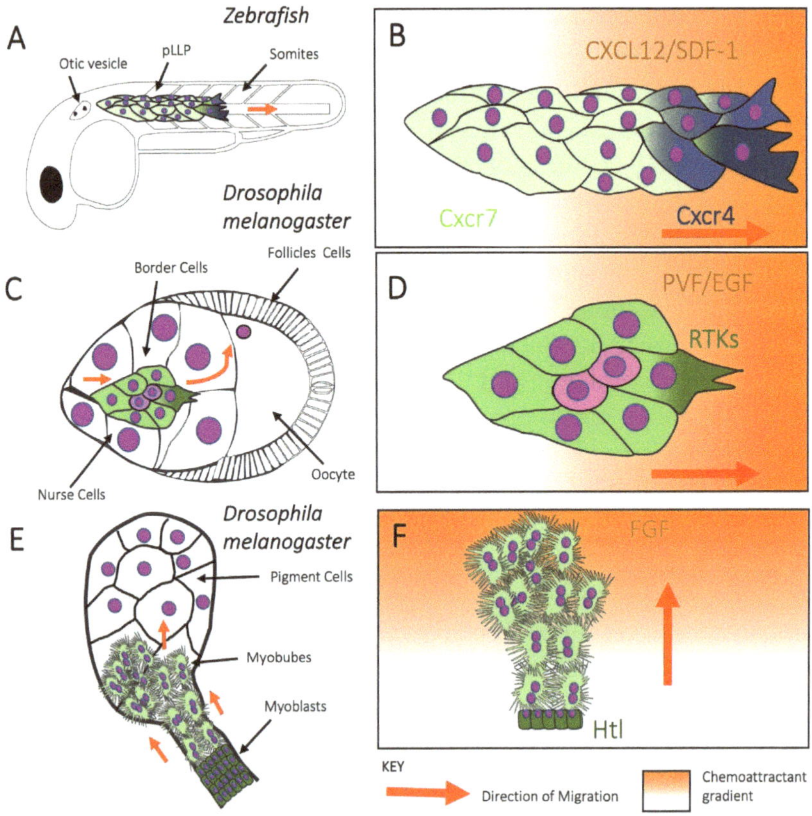

Fig. 4 Collective migration models during development. (A) In the zebrafish embryo, the lateral line primordium travels toward the tail along the horizontal myoseptum, located between the somites. The red arrow marks the migration path (B) The chemoattractant CXCL12/SDF-1 guides the movement of the primordium shown by the orange shading. The trailing cells express Cxcr7 (green), along with Cxcr4 receptors (though not visible in the figure). At the front, the leader cells prominently express Cxcr4 (blue) and form large protrusions that direct the migration. Stable cell–cell interactions are maintained throughout the primordium, ensuring coordinated movement. (C) In the *Drosophila* embryo, border cells detach from the anterior pole of the egg chamber and begin their migration toward the posterior, as shown by the straight red arrow. As they approach the end of their journey, they shift direction and turn dorsally, indicated by the curved red arrow. (D) The border cell cluster migrates through the nurse cells, maintaining strong cell-cell adhesions between the non-motile polar cells (purple) and the motile border cells (green). The migration is driven by a gradient of the chemoattractants PVR and EGF, shown by the orange shading. PVF-1 and EGF signaling polarize the protrusions in the border cells with the highest RTK activity, guiding the collective movement of the cluster. (E) In the developing *Drosophila* testes, myotubes, which are formed from the fusion of myoblasts, migrate collectively towards the testis tip contributing to the coiling of the testis for its final structure. The myoblasts,

(Fig. 4C and D). This process involves a specialized group of 6–10 follicular epithelial cells that detach from the anterior follicular epithelium and collectively navigate a path between the 15 nurse cells, covering approximately 150 μm until they reach the anterior border of the oocyte (Montell, Rorth, & Spradling, 1992; Wu, Tanwar, & Raftery, 2008). The migration occurs in two phases: an initial posteriorly directed migration towards the oocyte and a subsequent shorter dorsal movement, where the border cells move a short distance dorsally towards the germinal vesicle. Studies have identified a *Drosophila* ligand from the PDGF/VEGF family (PVF1) and its PDGF/VEGF Receptor (PVR), demonstrating their importance in the first migration phase. The dorsal movement is guided by EGF receptor (EGFR) signaling, which is not essential for the initial migration (Duchek & Rorth, 2001). The guidance function of PVR is largely redundant with that of EGFR, and both affect actin accumulation in follicle cells through the activation of downstream pathways involving Rac and Myoblast city (Mbc), which has been considered a non-conventional guanine nucleotide exchange factor (GEF) and acts as an activator of Rac, which is crucial for normal border cell migration (Duchek, Somogyi, Jékely, Beccari, & Rorth, 2001). Upon arrival at the oocyte, they attach to both the oocyte and the follicle cells that are moving inward. This process, called neo-lamination, helps to form a continuous epithelium to protect the oocyte (Miao, Godt, & Montell, 2020). This collaboration between border and follicle cells is essential for constructing the eggshell structure, vital for sperm entry and secretion of patterning signals whose failure results in female sterility (Montell et al., 1992; Montell, 2003; Prasad & Montell, 2007). Unlike other migratory clusters that rely on ECM interactions, border cells migrate exclusively in a cell-on-cell environment. Their movement depends on E-cadherin-mediated adhesion rather than integrin-ECM interactions. E-cadherin is essential for both cluster cohesion and signaling to Rac1, which promotes actin cytoskeletal remodeling and directional migration (Cai et al., 2014).

located at the base, give rise to the elongated myotubes that migrate through gaps between large pigment cells. Strong cell-cell adhesions coordinated by filopodia are maintained between the migrating myotubes to ensure coordinated movement. The red arrows indicate the direction of collective migration. (F) The migration of myotubes in the Drosophila testis is regulated by a gradient of FGF (indicated by the orange shading). Myoblasts express the Heartless (Htl) receptor, which is essential for detecting and responding to the FGF signal. Throughout this process, the cells maintain tight cell-cell adhesions to ensure coordinated collective movement. The direction of migration is shown by the red arrow.

Chemotactic cues from the oocyte and neighboring nurse cells play a pivotal role in precisely guiding border cells towards their target, ensuring timely arrival and correct positioning during oogenesis. Live imaging studies reveal dynamic behavior within the border cell cluster, highlighting the exchange of positions between leader and follower cells over time as they navigate through the egg chamber (Prasad & Montell, 2007). This model elucidates fundamental mechanisms underlying collective cell movement and cell-cell interactions, offering insights into tissue morphogenesis and directional migration regulation in development. The migration of border cells is orchestrated by receptor tyrosine kinase (RTK) signaling, where downstream effectors such as Rac and Cdc42 regulate actin cytoskeleton dynamics and cell polarity, critical for guiding migration towards the oocyte (Duchek & Rorth, 2001; Montell, 2003). Crucial cell-cell adhesion molecules like E-cadherin maintain cohesion within the migrating cell cluster, and are essential for signaling to Rac1, which promotes actin cytoskeletal remodeling and directional migration (Cai et al., 2014).

The distinct dynamics of leader and follower cells in epithelial collective migration are critical in this model. The acquisition and stability of leader positions over time are essential for effective migration and the formation of structural patterns within the developing egg chamber (Montell, 2003; Rorth, 2002). The collective migration of border cells is a common example of a small cluster with only one leader cell at a time. This leader cell extends forward in the direction of movement due to signals from chemoattractant receptors, which activate Rac and promote actin polymerization to create leading-edge extensions. The leader cell binds to follower cells via E-cadherin, thereby inhibiting them from generating protrusions. Recently, it has been discovered that follower cells also require Rac activity, but it is activated through a different pathway. It is hypothesized that this is because follower cells do not depend on chemoattractant receptor signaling to activate Rac. Instead, they need this GTPase activation to maintain cluster compactness and coordinated movement. Follower cells exhibit behaviors where they crawl on and over one another. Results show that the basolateral polarity proteins Scribble, Discs Large, and Lethal Giant Larva recruit a Rac activator to cell-cell contacts, which activates follower cell crawling behaviors (Campanale, Mondo, & Montell, 2022). These behaviors maintain cluster cohesion and keep the cluster moving, even when the lead cell is not protruding. All follower border cells are motile and can overtake leader cells. Thus, followers play multiple, critical roles in promoting efficient, collective border cell migration, including taking over the lead when necessary and maintaining coordinated collective migration in vivo (Campanale et al., 2022; Wang, He, Wu, Hahn, & Montell, 2010).

5.3 Drosophila testis myotubes

The migration of myotubes *Drosophila* testis is a good example of how groups of cells use filopodia in cell-cell contact to stimulate collective migration (Fig. 4E and F). During *Drosophila* pupal testis development, multinuclear testis myotubes migrate through gaps between large pigment cells and cyst cells. As they advance towards the testis tip, the myotubes start elongating, and eventually, the testis undergoes significant coiling to achieve its final structure (Bischoff, Lieb, Renkawitz-Pohl, & Bogdan, 2021; Kuckwa, Fritzen, Buttgereit, Rothenbusch-Fender, & Renkawitz-Pohl, 2016; Rothenbusch-Fender et al., 2017). This migration relies on filopodia from myoblasts that sense and respond to environmental cues, helping the myotubes elongate and organize to form the proper muscle structure. Because myotubes remain in continuous contact, individual cells cannot migrate independently, resulting in a collective movement. Myotubes extend filopodial protrusions in all directions, and these protrusions persist regardless of cell-cell contacts within the sheet. Although N-Cadherin adhesions are present, a localized reduction in matrix adhesion disrupts symmetry, leading to non-uniform protrusions. This reduction causes constant renewal of interdigitating filopodia-based adhesions. RhoA activity sporadically pulses along the open edges, likely responding to localized tension (Bischoff & Bogdan, 2021).

These mechanisms, referred to as Contact Stimulation of Migration (CSM), resemble the well-described Contact Inhibition of Locomotion (CIL) observed in *Xenopus* neural crest cells. Bischoff & Bogdan, (2021) highlight that while CSM and CIL share similarities, they function differently. CIL causes cells to repel each other upon contact, leading to directional changes to prevent crowding, and maintaining dispersion necessary for neural crest cell migration in *Xenopus*. In contrast, CSM promotes cohesive group movement through constant cell adhesion. Both processes showcase self-regulation: CSM through contact-dependent local inhibition of substrate adhesion, and CIL through contact-dependent repulsion and loss of protrusions, highlighting their roles in effective collective migration in different biological contexts (Bischoff & Bogdan, 2021).

The importance of chemotaxis in the collective migration of *Drosophila* testis myotubes lies in its regulation by the Heartless (Htl) receptor and its ligands, Thisbe (Ths) and Pyramus (Pyr). Htl, a fibroblast growth factor receptor (FGFR) expressed in myoblasts on the prospective seminal vesicles, is activated by Ths and Pyr, which are secreted from an unknown source. This activation is crucial for guiding the myoblasts as they migrate towards the testis to form a muscle sheath. Htl signaling promotes cytoskeletal rearrangements and the expression of non-muscle myosin II, essential for proper myotube migration and fusion. This chemotactic mechanism ensures coordinated collective migration, allowing the nascent myotubes to effectively cover and shape the testis, thereby stabilizing its structure during *Drosophila* metamorphosis (Bischoff et al., 2021; Rothenbusch-Fender et al., 2017).

6. Beyond collective cell migration: supracellular organization

Supracellular organization or migration refers to a subtype of collective migration where a group of cells behave as one single "supracell", increasing the scale of coordination compared to single-cell migration (Pfeifer, Shyer, & Rodrigues, 2024; Shellard & Mayor, 2019).

Rather than being distinct categories, supracellular migration can be considered a specialized form of collective migration, where cluster movement is better described at a level higher than the cells, where all the cells are integrated into a single coordinated unit (Fig. 5B). On the other hand, collective migration is based on the coordinated and cooperative movement of the individual cells within the cluster (Fig. 5A). During collective cell migration, an example is the epithelial cell migration, the individual cells that make up the cluster exhibit front-to-rear polarity, are connected by intercellular junctions for adhesion, equally contribute to the movement during migration, and sustain cryptic protrusions during movement. In supracellular migration cells do not exhibit front-to-rear polarity and there are no longer cryptic protrusions that form between cells (Shellard & Mayor, 2019; Venhuizen & Zegers, 2017). However, the entire cluster displays an overall front-to-rear polarity, whereby leader cells become the "front" and the rear of the cells in the cluster periphery as the "rear". This suggests a spectrum of collective migration behaviors, where supracellular organization represents a high degree of coordination at the tissue or cluster level rather than a

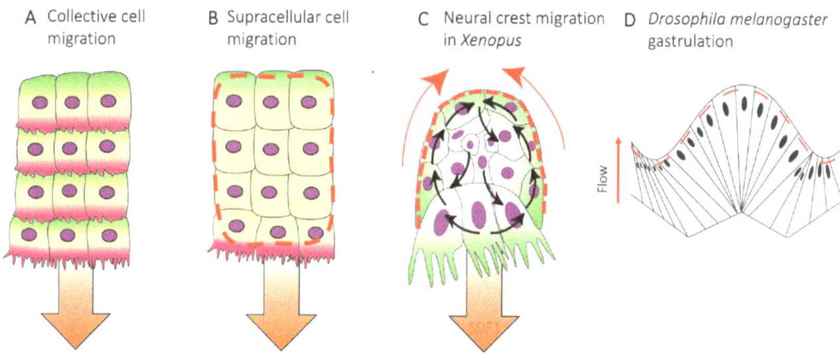

Fig. 5 Examples of supracellular organization in early development. (A) Collective cell migration: The cells are linked through intercellular adhesions in migrating cell clusters. Each cell has a defined front (pink) and rear (green) and plays an equal role in the group's movement. Every cell also keeps its protrusions, which help facilitate the movement. (B) Supracellular migration: The cell group behaves as a single entity. The individual cells lose their distinct front-rear orientation, and their connections disappear. Instead, the entire group displays a collective front-rear polarity, where leader cells at the front (pink) and cells at the rear (green) guide the overall movement. Actomyosin cable (red dashed line) is one of the characteristic features in supracellular migration and plays a crucial role in maintaining cohesion and coordinated movement. (C) In *Xenopus*, the neural crest is a prominent example that displays behaviors of a "supracell". In the multicellular cluster, there is a clear front-to-rear polarity, and the peripheral cells are linked by a tensile actomyosin cable (red dashed line). Cellular protrusions are observed in the outer cells on the leading edge of the cluster. The neural crest undergoes directional migration towards the gradient of SDF1, a chemoattractant. (D) In *Drosophila*, ventral furrow formation during gastrulation is another example of supracellular organization, where sheets of epithelial cells fold. Myosin II is phosphorylated at the apical section of the cells (red dashed line), which initiates anisotropic constriction along with the flow of cells.

fundamentally separate mode of migration. For an extensive description and examples of supracellular migration see (Shellard & Mayor, 2019). This section of the review will focus on specific in vivo examples of migration in neural crest in *Xenopus* and in gastrulation in *Drosophila melanogaster*.

6.1 Neural crest migration in *Xenopus*

The neural crest is a transient embryonic cell population, which migrates to distant target sites where they differentiate into many different cell types. During the early development of *Xenopus* embryos, neural crest cells migrate collectively towards the chemoattractant SDF-1, and the supracellular cluster is polarized in the direction of chemotaxis (Shellard, Szabó, Trepat, & Mayor, 2018; Theveneau et al., 2010). One of the mechanisms that explain the supracellular polarity of neural crest cell clusters during chemotaxis is contact inhibition of locomotion (CIL), where migrating cells come into contact with each other and repolarize away from each other (Carmona-Fontaine et al., 2008; Stramer & Mayor, 2017; Yoon et al., 2018). Within the cluster cell protrusions are absent, and protrusions are only observed in the outer leading cells as a result of CIL (Carmona-Fontaine et al., 2008). Thus, an overall front-to-rear polarity is established for the supracellular cluster (Fig. 5C). Additionally, cells at the periphery of the cluster are linked by a tensile actomyosin cable, which supports the supracellular organization of cytoskeletal components within the cluster (Shellard et al., 2018). Furthermore, it has been shown that traction forces were observed in cells at the periphery of the cluster compared to the absence of these traction forces in cells at the core of the cluster, thereby maintaining supracellular polarity of the cluster (Roycroft et al., 2018; Scarpa et al., 2015) However, it is unclear whether the tractions forces are transmitted through the supracellular cluster. In addition to traction forces, when the actomyosin cable at the rear of the supracellular cluster contracts during chemotaxis towards SDF-1, the cells at the core of the cluster move forward and the cells in the outer part of the cluster move backwards (to the rear), thereby coordinating the forward movement of the cluster to the chemoattractant.

6.2 Drosophila melanogaster gastrulation

An example where there is strong evidence for supracellular behavior is gastrulation in *Drosophila*, as it has been shown that individual cells are not strictly necessary for the overall migration of the tissue. During *Drosophila* gastrulation, sheets of epithelial cells fold during the ventral furrow formation (Fig. 5D). This process begins when myosin II is phosphorylated at the apical part of the cells, which initiates anisotropic constriction that leads to tissue invagination (inward movement of cells) (He, Doubrovinski, Polyakov, & Wieschaus, 2014; Martin & Goldstein, 2014; Martin, Gelbart, Fernandez-Gonzalez, Kaschube, & Wieschaus, 2010; Monier et al., 2015). The movement of the tissue is analogous to the behavior of a viscous liquid, where subcellular components do not impact the supracellular organization of the cells (He et al., 2014). Furthermore, specific spatio-temporal activation of Rho signaling can induce apical constriction and epithelial folding, and they can occur at any orientation of the embryo axis (Izquierdo, Quinkler, & De Renzis, 2018). It has been shown that in mutant embryos that cannot form individual cells at the blastula stage ("acellular" embryos), apical constriction at the equivalent gastrula stages resulted in similar fluid flow patterns compared to the wild-type embryos (He et al., 2014). The same study also concluded that the cell membrane had a passive flow and did not cause any driving force in the invagination nor any resistance to the movement (He et al., 2014). Whilst there is an open question regarding any changes to cellular volume during ventral furrow formation, a computational vertex model reported that the volume of the apically constricting cells is maintained throughout invagination and cell shape changes are regulated by the stiffness of the lateral and basal membrane surfaces (Polyakov et al., 2014). Therefore, these observations emphasize that ventral furrow formation is a supracellular phenomenon, where the invaginating epithelial cells behave as a "supracell".

7. Concluding remarks

The mechanisms of cell migration across various biological models reviewed here underscore a crucial insight: substantial cooperativity and association are frequently present when cells migrate collectively and suggest that a similar cell-cell interaction takes place in what traditionally be considered single cell migration. This cooperativity is essential for effective migration and function, ensuring proper tissue formation and adaptation to environmental cues, and reflects fundamental principles of developmental biology.

Single-cell migration, such as primordial germ cell migration in zebrafish and *Drosophila melanogaster* reveals how individual cells are meticulously guided by chemoattractant cues and ECM interactions. These processes ensure precise positioning during early development, highlighting that even seemingly independent migrations involve significant coordination and responsiveness to external signals. Whilst primordial germ migration is considered a single-cell migration phenomenon in vivo, it is unclear whether individual cells are not showing any kind of cooperative migration behavior. One aspect of primordial germ cell migration is the directionality controlled by chemotaxis. We discussed that zebrafish utilize SDF-1A as a chemoattractant, whereas *Drosophila*, utilize phospholipids. These observations pose the question of whether cooperativity is working between cells when they respond to chemotactic cues, despite the lack of cell-cell connections; for example, the internalization of ligands by a cell could contribute to the generation of chemoattractant gradient that will affect the chemotaxis of its neighbors. Future studies will need to address whether migrating cells can shift between cooperative or supracellular migration and single-cell migration depending on the mechanical or chemical stimulus.

Collective migration, exemplified by the lateral line primordium in *zebrafish*, is orchestrated by leader and follower cells. Leader cells sense and navigate external signals, while follower cells maintain cohesion, force generation and shape the chemoattract gradient. This interplay is crucial for the successful colonization of target sites. Additionally, this model showcases mesenchymal migration, characterized by elongated cells moving through the extracellular matrix using integrin-mediated adhesions and actin polymerization. In the context of collective cell migration, actomyosin cables play a crucial role in maintaining cohesion and enabling coordinated movement, as previously reviewed by Bischoff and colleagues (Bischoff & Bogdan, 2021). These cables are essential for generating the mechanical forces needed for supracellular organization, a state where cell clusters move as a cohesive unit. In *Drosophila* border cell clusters and cephalic neural crest in *Xenopus*, peripheral myosin activity creates the cortical tension necessary to resist the external pressure from surrounding nurse cells. This tension is maintained by a network of supracellular actomyosin cables that surround the cluster and enforce cohesion, preventing follower cells from forming protrusions This structure ensures that the cluster moves as a single, cohesive unit, demonstrating the importance of actomyosin cables in maintaining the integrity and directionality of migrating cells (Aranjuez, Burtscher, Sawant, Majumder, & McDonald, 2016; Shellard et al., 2018; Wang, Guo, Wang, Wang, & Chen, 2020). This mechanism highlights the critical role of actomyosin cables in facilitating effective communication and force transmission across the cluster, enabling the cells to migrate efficiently in response to chemotactic cues. Migrating testis myotubes in *Drosophila* utilize actomyosin cables differently, as they close internal gaps like wound healing or dorsal closure. This internal cable formation is crucial for maintaining the structural integrity of the myotubes as they migrate to cover and shape the testis. By closing these gaps, actomyosin cables ensure that the myotubes maintain a cohesive structure, enabling them to move as a unified entity and effectively reshape the testis (Rothenbusch-Fender et al., 2017).

The generation of actomyosin cables in these various models underscores their importance in achieving supracellular organization. Actomyosin cables provide the tensile strength and mechanical stability necessary for maintaining cohesion within migrating cell clusters. They facilitate the transmission of forces across cells, ensuring that the entire cluster moves in a coordinated and directed manner. This supracellular organization is essential for effective collective migration, allowing cells to respond to environmental cues and navigate through complex tissue environments. Additionally, by creating a contractile network that integrates mechanical and chemical signals, actomyosin cables coordinate the behavior of individual cells within the cluster. This coordination is essential for maintaining the directionality and efficiency of migration, ensuring that the cells can achieve their developmental and morphogenetic goals. Actomyosin cables are indispensable for achieving supracellular organization in collective cell migration. Whether maintaining peripheral tension in neural crest cells and border cell clusters or closing internal gaps in migrating myotubes, actomyosin cables provide the structural framework necessary for cohesive and coordinated movement. Understanding the molecular and mechanical bases of these cables is crucial for insights into developmental biology and the treatment of pathological conditions involving aberrant cell migration.

Acknowledgments

We thank Joseph Campanale for his comments on the manuscript. Work in R.M. laboratory is supported by grants from the Medical Research Council (MR/S007792/1), Biotechnology and Biological Sciences Research Council (M008517, BB/T013044), and Wellcome Trust (102489/Z/13/Z). The authors declare no conflict of interest.

References

Alivernini, S., Firestein, G. S., & McInnes, I. B. (2022). The pathogenesis of rheumatoid arthritis. *Immunity, 55*(12), 2255–2270. https://doi.org/10.1016/j.immuni.2022.11.009.

Alvizi, L., Nani, D., Brito, L. A., Kobayashi, G. S., Passos-Bueno, M. R., & Mayor, R. (2023). Neural crest E-cadherin loss drives cleft lip/palate by epigenetic modulation via pro-inflammatory gene–environment interaction. *Nature Communications, 14*(1), 2868. https://doi.org/10.1038/s41467-023-38526-1.

Aman, A., & Piotrowski, T. (2010). Cell migration during morphogenesis. *Developmental Biology, 341*(1), 20–33. https://doi.org/10.1016/j.ydbio.2009.11.014.

Aranjuez, G., Burtscher, A., Sawant, K., Majumder, P., & McDonald, J. A. (2016). Dynamic myosin activation promotes collective morphology and migration by locally balancing oppositional forces from surrounding tissue. *Molecular Biology of the Cell, 27*(12), https://doi.org/10.1091/mbc.E15-10-0744.

Barton, L. J., Roa-De La Cruz, L., Lehmann, R., & Lin, B. (2024). The journey of a generation: Advances and promises in the study of primordial germ cell migration. *Development (Cambridge, England), 151*(7), dev201102. https://doi.org/10.1242/dev.201102.

Bischoff, M. C., & Bogdan, S. (2021). Collective cell migration driven by filopodia-New insights from the social behavior of myotubes. *Bioessays: News and Reviews in Molecular, Cellular and Developmental Biology, 43*(11), https://doi.org/10.1002/bies.202100124.

Bischoff, M. C., Lieb, S., Renkawitz-Pohl, R., & Bogdan, S. (2021). Filopodia-based contact stimulation of cell migration drives tissue morphogenesis. *Nature Communications, 12*(1), https://doi.org/10.1038/s41467-020-20362-2.

Blaser, H., Eisenbeiss, S., Neumann, M., Reichman-Fried, M., Thisse, B., Thisse, C., & Raz, E. (2005). Transition from non-motile behaviour to directed migration during early PGC development in zebrafish. *Journal of Cell Science, 118*(17), 4027–4038. https://doi.org/10.1242/jcs.02522.

Blaser, H., Reichman-Fried, M., Castanon, I., Dumstrei, K., Marlow, F. L., Kawakami, K., ... Raz, E. (2006). Migration of zebrafish primordial germ cells: A role for myosin contraction and cytoplasmic flow. *Developmental Cell, 11*(5), 613–627. https://doi.org/10.1016/j.devcel.2006.09.023.

Boldajipour, B., Mahabaleshwar, H., Kardash, E., Reichman-Fried, M., Blaser, H., Minina, S., ... Raz, E. (2008). Control of chemokine-guided cell migration by ligand sequestration. *Cell, 132*(3), 463–473. https://doi.org/10.1016/j.cell.2007.12.034.

Bussmann, J., & Raz, E. (2015). Chemokine-guided cell migration and motility in zebrafish development. *The EMBO Journal, 34*(10), 1309–1318. https://doi.org/10.15252/embj.201490105.

Cai, D., Chen, S. C., Prasad, M., He, L., Wang, X., Choesmel-Cadamuro, V., ... Montell, D. J. (2014). Mechanical feedback through E-cadherin promotes direction sensing during collective cell migration. *Cell, 157*(5), 1146–1159. https://doi.org/10.1016/j.cell.2014.03.045.

Campanale, J. P., Mondo, J. A., & Montell, D. J. (2022). A Scribble/Cdep/Rac pathway controls follower-cell crawling and cluster cohesion during collective border-cell migration. *Developmental Cell, 57*(21), 2483–2496.e4. https://doi.org/10.1016/j.devcel.2022.10.004.

Campanale, J. P., & Montell, D. J. (2023). Who's really in charge: Diverse follower cell behaviors in collective cell migration. *Current Opinion in Cell Biology, 81*, 102160. https://doi.org/10.1016/j.ceb.2023.102160.

Carmona-Fontaine, C., Matthews, H. K., Kuriyama, S., Moreno, M., Dunn, G. A., Parsons, M., ... Mayor, R. (2008). Contact inhibition of locomotion in vivo controls neural crest directional migration. *Nature, 456*(7224), 957–961. https://doi.org/10.1038/nature07441.

Charras, G., & Paluch, E. (2008). Blebs lead the way: How to migrate without lamellipodia. *Nature Reviews. Molecular Cell Biology, 9*(9), 730–736. https://doi.org/10.1038/nrm2453.

Cubedo, N., Cerdan, E., Sapede, D., & Rossel, M. (2009). CXCR4 and CXCR7 cooperate during tangential migration of facial motoneurons. *Molecular and Cellular Neuroscience, 40*(4), 474–484. https://doi.org/10.1016/j.mcn.2009.01.003.

Doitsidou, M., Reichman-Fried, M., Stebler, J., Köprunner, M., Dörries, J., Meyer, D., ... Raz, E. (2002). Guidance of primordial germ cell migration by the chemokine SDF-1. *Cell, 111*(5), 647–659. https://doi.org/10.1016/S0092-8674(02)01135-2.

Donà, E., Barry, J. D., Valentin, G., Quirin, C., Khmelinskii, A., Kunze, A., ... Gilmour, D. (2013). Directional tissue migration through a self-generated chemokine gradient. *Nature, 503*(7475), 285–289. https://doi.org/10.1038/nature12635.

Duchek, P., & Rorth, P. (2001). Guidance of cell migration by EGF receptor signaling during Drosophila oogenesis. *Science (New York, N. Y.), 291*(5501), https://doi.org/10.1126/science.291.5501.131.

Duchek, P., Somogyi, K., Jékely, G., Beccari, S., & Rorth, P. (2001). Guidance of cell migration by the Drosophila PDGF/VEGF receptor. *Cell, 107*(1), https://doi.org/10.1016/S0092-8674(01)00502-5.

Engelhardt, B., & Liebner, S. (2014). Novel insights into the development and maintenance of the blood–brain barrier. *Cell and Tissue Research, 355*(3), 687–699. https://doi.org/10.1007/s00441-014-1811-2.

Farooqui, R., & Fenteany, G. (2005). Multiple rows of cells behind an epithelial wound edge extend cryptic lamellipodia to collectively drive cell-sheet movement. *Journal of Cell Science, 118*(1), https://doi.org/10.1242/jcs.01577.

Friedl, P., & Gilmour, D. (2009). Collective cell migration in morphogenesis, regeneration and cancer. *Nature Reviews. Molecular Cell Biology, 10*(7), 445–457. https://doi.org/10.1038/nrm2720.
Friedl, P., & Wolf, K. (2010). Plasticity of cell migration: A multiscale tuning model. *Journal of Cell Biology, 188*(1), 11–19. https://doi.org/10.1083/jcb.200909003.
García-Arcos, J. M., Jha, A., Waterman, C. M., & Piel, M. (2024). Blebology: Principles of bleb-based migration. *Trends in Cell Biology*, S0962892424000321. https://doi.org/10.1016/j.tcb.2024.02.009.
Gerhart, J., & Keller, R. (1986). Region-specific cell activities in amphibian gastrulation (Annual Reviews) *Annual Review of Cell and Developmental Biology, 2*, 201–229. https://doi.org/10.1146/annurev.cb.02.110186.001221.
Ghysen, A., & Dambly-Chaudière, C. (2004). Development of the zebrafish lateral line. *Current Opinion in Neurobiology, 14*(1), https://doi.org/10.1016/j.conb.2004.01.012.
Gimbrone, M. A., & García-Cardeña, G. (2016). Endothelial cell dysfunction and the pathobiology of atherosclerosis. *Circulation Research, 118*(4), 620–636. https://doi.org/10.1161/CIRCRESAHA.115.306301.
Goudarzi, M., Banisch, T. U., Mobin, M. B., Maghelli, N., Tarbashevich, K., Strate, I., ... Raz, E. (2012). Identification and regulation of a molecular module for bleb-based cell motility. *Developmental Cell, 23*(1), 210–218. https://doi.org/10.1016/j.devcel.2012.05.007.
Guillot, C., & Lecuit, T. (2013). Mechanics of epithelial tissue homeostasis and morphogenesis. *Science (New York, N. Y.), 340*(6137), https://doi.org/10.1126/science.1235249.
Haas, P., & Gilmour, D. (2006). Chemokine signaling mediates self-organizing tissue migration in the zebrafish lateral line. *Developmental Cell, 10*(5), https://doi.org/10.1016/j.devcel.2006.02.019.
Hanahan, D. (2022). Hallmarks of cancer: New dimensions. *Cancer Discovery, 12*(1), 31–46. https://doi.org/10.1158/2159-8290.CD-21-1059.
He, B., Doubrovinski, K., Polyakov, O., & Wieschaus, E. (2014). Apical constriction drives tissue-scale hydrodynamic flow to mediate cell elongation. *Nature, 508*(7496), 392–396. https://doi.org/10.1038/nature13070.
Hynes, R. O. (2002). Integrins: Bidirectional, allosteric signaling machines. *Cell, 110*(6), https://doi.org/10.1016/S0092-8674(02)00971-6.
Hynes, R. O. (2009). The extracellular matrix: Not just pretty fibrils. *Science (New York, N. Y.), 326*(5957), 1216–1219. https://doi.org/10.1126/science.1176009.
Izquierdo, E., Quinkler, T., & De Renzis, S. (2018). Guided morphogenesis through optogenetic activation of Rho signalling during early Drosophila embryogenesis. *Nature Communications, 9*(1), 2366. https://doi.org/10.1038/s41467-018-04754-z.
Jaglarz, M. K., & Howard, K. R. (1994). Primordial germ cell migration in *Drosophila melanogaster* is controlled by somatic tissue. *Development (Cambridge, England), 120*(1), 83–89. https://doi.org/10.1242/dev.120.1.83.
Jaglarz, M. K., & Howard, K. R. (1995). The active migration of *Drosophila* primordial germ cells. *Development (Cambridge, England), 121*(11), 3495–3503. https://doi.org/10.1242/dev.121.11.3495.
Joyce, J. A., & Pollard, J. W. (2009). Microenvironmental regulation of metastasis. *Nature Reviews. Cancer, 9*(4), 239–252. https://doi.org/10.1038/nrc2618.
Kalluri, R., & Weinberg, R. A. (2009). The basics of epithelial-mesenchymal transition. *Journal of Clinical Investigation, 119*(6), 1420–1428. https://doi.org/10.1172/JCI39104.1420.
Kalluri, R., & Weinberg, R. A. (2010). Erratum: The basics of epithelial-mesenchymal transition (The Journal of Clinical Investigation (2009) 119, 6, (1420–1428) https://doi.org/10.1172/JCI39104). *Journal of Clinical Investigation, 120*(5). https://doi.org/10.1172/JCI39104C1.
Kardash, E., Reichman-Fried, M., Maître, J.-L., Boldajipour, B., Papusheva, E., Messerschmidt, E.-M., ... Raz, E. (2010). A role for Rho GTPases and cell–cell adhesion in single-cell motility in vivo. *Nature Cell Biology, 12*(1), 47–53. https://doi.org/10.1038/ncb2003.

Kenwrick, K., Mukherjee, A., & Renault, A. D. (2019). Hmgcr promotes a long-range signal to attract *Drosophila* germ cells independently of Hedgehog. *Journal of Cell Science, 132*(23), jcs232637. https://doi.org/10.1242/jcs.232637.

Knaut, H., Pelegri, F., Bohmann, K., Schwarz, H., & Nüsslein-Volhard, C. (2000). Zebrafish *vasa* RNA but not its protein is a component of the germ plasm and segregates asymmetrically before germline specification. *The Journal of Cell Biology, 149*(4), 875–888. https://doi.org/10.1083/jcb.149.4.875.

Köprunner, M., Thisse, C., Thisse, B., & Raz, E. (2001). A zebrafish *nanos* -related gene is essential for the development of primordial germ cells. *Genes & Development, 15*(21), 2877–2885. https://doi.org/10.1101/gad.212401.

Kuckwa, J., Fritzen, K., Buttgereit, D., Rothenbusch-Fender, S., & Renkawitz-Pohl, R. (2016). A new level of plasticity: Drosophila smooth-like testes muscles compensate failure of myoblast fusion. *Development (Cambridge), 143*(2), https://doi.org/10.1242/dev.126730.

Kunwar, P. S., Sano, H., Renault, A. D., Barbosa, V., Fuse, N., & Lehmann, R. (2008). Tre1 GPCR initiates germ cell transepithelial migration by regulating *Drosophila melanogaster* E-cadherin. *The Journal of Cell Biology, 183*(1), 157–168. https://doi.org/10.1083/jcb.200807049.

Lämmermann, T., & Sixt, M. (2009). Mechanical modes of 'amoeboid' cell migration. *Current Opinion in Cell Biology, 21*(5), 636–644. https://doi.org/10.1016/j.ceb.2009.05.003.

Lau, S., Feitzinger, A., Venkiteswaran, G., Wang, J., Lewellis, S. W., Koplinski, C. A., ... Knaut, H. (2020). A negative-feedback loop maintains optimal chemokine concentrations for directional cell migration. *Nature Cell Biology, 22*(3), https://doi.org/10.1038/s41556-020-0465-4.

Lin, B., Luo, J., & Lehmann, R. (2020). Collectively stabilizing and orienting posterior migratory forces disperses cell clusters in vivo. *Nature Communications, 11*(1), 4477. https://doi.org/10.1038/s41467-020-18185-2.

Lin, B., Luo, J., & Lehmann, R. (2022). An AMPK phosphoregulated RhoGEF feedback loop tunes cortical flow–driven amoeboid migration in vivo. *Science Advances, 8*(37), eabo0323. https://doi.org/10.1126/sciadv.abo0323.

Liu, Y. J., Le Berre, M., Lautenschlaeger, F., Maiuri, P., Callan-Jones, A., Heuzé, M., ... Piel, M. (2015). Confinement and low adhesion induce fast amoeboid migration of slow mesenchymal cells. *Cell, 160*(4), 659–672. https://doi.org/10.1016/j.cell.2015.01.007.

Mahabaleshwar, H., Tarbashevich, K., Nowak, M., Brand, M., & Raz, E. (2012). β-Arrestin control of late endosomal sorting facilitates decoy receptor function and chemokine gradient formation. *Development (Cambridge, England), 139*(16), 2897–2902. https://doi.org/10.1242/dev.080408.

Mantovani, A., Allavena, P., Sica, A., & Balkwill, F. (2008). Cancer-related inflammation. *Nature, 454*(7203), 436–444. https://doi.org/10.1038/nature07205.

Martin, A. C., Gelbart, M., Fernandez-Gonzalez, R., Kaschube, M., & Wieschaus, E. F. (2010). Integration of contractile forces during tissue invagination. *Journal of Cell Biology, 188*(5), 735–749. https://doi.org/10.1083/jcb.200910099.

Martin, A. C., & Goldstein, B. (2014). Apical constriction: Themes and variations on a cellular mechanism driving morphogenesis. *Development (Cambridge, England), 141*(10), 1987–1998. https://doi.org/10.1242/dev.102228.

Massimo De, F., Scaldaferri, M. L., & Farini, D. (2005). Adhesion molecules for mouse primordial germ cells. *Frontiers in Bioscience, 10*(1), 542–551. https://doi.org/10.2741/1550.

Mayor, R., & Theveneau, E. (2013). The neural crest. *Development (Cambridge, England), 140*(11), 2247–2251. https://doi.org/10.1242/dev.091751.

Miao, G., Godt, D., & Montell, D. J. (2020). Integration of migratory cells into a new site in vivo requires channel-independent functions of innexins on microtubules. *Developmental Cell, 54*(4), 501–515.e9. https://doi.org/10.1016/j.devcel.2020.06.024.

Mishra, A. K., Campanale, J. P., Mondo, J. A., & Montell, D. J. (2019). Cell interactions in collective cell migration. *Development (Cambridge, England), 146*, 1–7. https://doi.org/10.1242/dev.172056.

Monier, B., Gettings, M., Gay, G., Mangeat, T., Schott, S., Guarner, A., & Suzanne, M. (2015). Apico-basal forces exerted by apoptotic cells drive epithelium folding. *Nature, 518*(7538), 245–248. https://doi.org/10.1038/nature14152.

Montell, D. J. (2003). Border-cell migration: The race is on. *Nature Reviews. Molecular Cell Biology, 4*(1), https://doi.org/10.1038/nrm1006.

Montell, D. J., Rorth, P., & Spradling, A. C. (1992). Slow border cells, a locus required for a developmentally regulated cell migration during oogenesis, encodes Drosophila C EBP. *Cell, 71*(1), https://doi.org/10.1016/0092-8674(92)90265-E.

Nobes, C. D., & Hall, A. (1995). Rho, Rac, and Cdc42 GTPases regulate the assembly of multimolecular focal complexes associated with actin stress fibers, lamellipodia, and filopodia. *Cell, 81*(1), 53–62. https://doi.org/10.1016/0092-8674(95)90370-4.

Olguin-Olguin, A., Aalto, A., Maugis, B., Boquet-Pujadas, A., Hoffmann, D., Ermlich, L., ... Raz, E. (2021). Chemokine-biased robust self-organizing polarization of migrating cells in vivo. *Proceedings of the National Academy of Sciences, 118*(7), e2018480118. https://doi.org/10.1073/pnas.2018480118.

Paluch, E. K., & Raz, E. (2013). The role and regulation of blebs in cell migration. *Current Opinion in Cell Biology, 25*(5), https://doi.org/10.1016/j.ceb.2013.05.005.

Paňková, K., Rösel, D., Novotný, M., & Brábek, J. (2010). The molecular mechanisms of transition between mesenchymal and amoeboid invasiveness in tumor cells. *Cellular and Molecular Life Sciences, 67*(1), 63–71. https://doi.org/10.1007/s00018-009-0132-1.

Pérez-González, C., Ceada, G., Greco, F., Matejčić, M., Gómez-González, M., Castro, N., ... Trepat, X. (2021). Mechanical compartmentalization of the intestinal organoid enables crypt folding and collective cell migration. *Nature Cell Biology, 23*(7), 745–757. https://doi.org/10.1038/s41556-021-00699-6.

Pfeifer, C. R., Shyer, A. E., & Rodrigues, A. R. (2024). Creative processes during vertebrate organ morphogenesis: Biophysical self-organization at the supracellular scale. *Current Opinion in Cell Biology, 86*, 102305. https://doi.org/10.1016/j.ceb.2023.102305.

Polyakov, O., He, B., Swan, M., Shaevitz, J. W., Kaschube, M., & Wieschaus, E. (2014). Passive mechanical forces control cell-shape change during drosophila ventral furrow formation. *Biophysical Journal, 107*(4), 998–1010. https://doi.org/10.1016/j.bpj.2014.07.013.

Prasad, M., & Montell, D. J. (2007). Cellular and molecular mechanisms of border cell migration analyzed using time-lapse live-cell imaging. *Developmental Cell, 12*(6), https://doi.org/10.1016/j.devcel.2007.03.021.

Randolph, G. J., Angeli, V., & Swartz, M. A. (2005). Dendritic-cell trafficking to lymph nodes through lymphatic vessels. *Nature Reviews. Immunology, 5*(8), 617–628. https://doi.org/10.1038/nri1670.

Raz, E. (2004). Guidance of primordial germ cell migration. *Current Opinion in Cell Biology, 16*(2), 169–173. https://doi.org/10.1016/j.ceb.2004.01.004.

Reichman-Fried, M., Minina, S., & Raz, E. (2004). Autonomous modes of behavior in primordial germ cell migration. *Developmental Cell, 6*(4), 589–596. https://doi.org/10.1016/S1534-5807(04)00074-7.

Ricardo, S., & Lehmann, R. (2009). An ABC transporter controls export of a *Drosophila* germ cell attractant. *Science (New York, N. Y.), 323*(5916), 943–946. https://doi.org/10.1126/science.1166239.

Richardson, B. E., & Lehmann, R. (2010). Mechanisms guiding primordial germ cell migration: Strategies from different organisms. *Nature Reviews. Molecular Cell Biology, 11*(1), 37–49. https://doi.org/10.1038/nrm2815.

Ridley, A. J., Paterson, H. F., Johnston, C. L., Diekmann, D., & Hall, A. (1992). The small GTP-binding protein rac regulates growth factor-induced membrane ruffling. *Cell, 70*(3), 401–410. https://doi.org/10.1016/0092-8674(92)90164-8.
Ridley, A. J., Schwartz, M. A., Burridge, K., Firtel, R. A., Ginsberg, M. H., Borisy, G., ... Horwitz, A. R. (2003). Cell migration: Integrating signals from front to back. *Science (New York, N. Y.), 302*(5651), 1704–1709. https://doi.org/10.1126/science.1092053.
Rorth, P. (2002). Initiating and guiding migration: Lessons from border cells. *Trends in Cell Biology, 12*(7), https://doi.org/10.1016/S0962-8924(02)02311-5.
Rørth, P. (2009). Collective cell migration. *Annual Review of Cell and Developmental Biology Vol. 25*, Annual Reviews, 407–429. https://doi.org/10.1146/annurev.cellbio.042308.113231.
Rørth, P. (2011). Whence directionality: Guidance mechanisms in solitary and collective cell migration. *Developmental Cell, 20*(1), 9–18. https://doi.org/10.1016/j.devcel.2010.12.014.
Rothenbusch-Fender, S., Fritzen, K., Bischoff, M. C., Buttgereit, D., Oenel, S. F., & Renkawitz-Pohl, R. (2017). Myotube migration to cover and shape the testis of Drosophila depends on Heartless, Cadherin/Catenin, and myosin II. *Biology Open, 6*(12), https://doi.org/10.1242/bio.025940.
Roycroft, A., Szabó, A., Bahm, I., Daly, L., Charras, G., Parsons, M., & Mayor, R. (2018). Redistribution of adhesive forces through Src/FAK drives contact inhibition of locomotion in neural crest. *Developmental Cell, 45*(5), 565–579.e3. https://doi.org/10.1016/j.devcel.2018.05.003.
Ruprecht, V., Wieser, S., Callan-Jones, A., Smutny, M., Morita, H., Sako, K., ... Heisenberg, C.-P. (2015). Cortical contractility triggers a stochastic switch to fast amoeboid cell motility. *Cell, 160*(4), 673–685. https://doi.org/10.1016/j.cell.2015.01.008.
Sahai, E., & Marshall, C. J. (2003). Differing modes of tumour cell invasion have distinct requirements for Rho/ROCK signalling and extracellular proteolysis. *Nature Cell Biology, 5*(8), 711–719. https://doi.org/10.1038/ncb1019.
Sano, H., Renault, A. D., & Lehmann, R. (2005). Control of lateral migration and germ cell elimination by the *Drosophila melanogaster* lipid phosphate phosphatases Wunen and Wunen 2. *The Journal of Cell Biology, 171*(4), 675–683. https://doi.org/10.1083/jcb.200506038.
Santos, A. C., & Lehmann, R. (2004). Germ cell specification and migration in Drosophila and beyond. *Current Biology, 14*(14), R578–R589. https://doi.org/10.1016/j.cub.2004.07.018.
Saraiva, J. E., & Barriga, E. H. (2021). The basics of collective cell migration: Unity makes strength. *Viscoelasticity and Collective Cell Migration: An Interdisciplinary Perspective Across Levels of Organization*. https://doi.org/10.1016/B978-0-12-820310-1.00001-X.
Scarpa, E., & Mayor, R. (2016). Collective cell migration in development. *Journal of Cell Biology, 212*(2), https://doi.org/10.1083/jcb.201508047.
Scarpa, E., Szabó, A., Bibonne, A., Theveneau, E., Parsons, M., & Mayor, R. (2015). Cadherin switch during EMT in neural crest cells leads to contact inhibition of locomotion via repolarization of forces. *Developmental Cell, 34*(4), 421–434. https://doi.org/10.1016/j.devcel.2015.06.012.
Schick, J., & Raz, E. (2022). Blebs—Formation, regulation, positioning, and role in amoeboid cell migration. *Frontiers in Cell and Developmental Biology, 10*, 926394. https://doi.org/10.3389/fcell.2022.926394.
Schumacher, L. (2019). Collective cell migration in development. *Advances in Experimental Medicine and Biology, 1146*. https://doi.org/10.1007/978-3-030-17593-1_7.
Shellard, A., & Mayor, R. (2019). Supracellular migration – Beyond collective cell migration. *Journal of Cell Science, 132*(8), jcs226142. https://doi.org/10.1242/jcs.226142.
Shellard, A., Szabó, A., Trepat, X., & Mayor, R. (2018). Supracellular contraction at the rear of neural crest cell groups drives collective chemotaxis. *Science (New York, N. Y.), 362*(6412), 339–343. https://doi.org/10.1126/science.aau3301.
Sofroniew, M. V. (2015). Astrocyte barriers to neurotoxic inflammation. *Nature Reviews. Neuroscience, 16*(5), 249–263. https://doi.org/10.1038/nrn3898.

Stramer, B., & Mayor, R. (2017). Mechanisms and in vivo functions of contact inhibition of locomotion. *Nature Reviews. Molecular Cell Biology, 18*(1), 43–55. https://doi.org/10.1038/nrm.2016.118.

Theveneau, E., Marchant, L., Kuriyama, S., Gull, M., Moepps, B., Parsons, M., & Mayor, R. (2010). Collective chemotaxis requires contact-dependent cell polarity. *Developmental Cell, 19*(1), 39–53. https://doi.org/10.1016/j.devcel.2010.06.012.

Thiery, J. P., Acloque, H., Huang, R. Y. J., & Nieto, M. A. (2009). Epithelial-mesenchymal transitions in development and disease. *Cell, 139*(5), https://doi.org/10.1016/j.cell.2009.11.007.

Trcek, T., & Lehmann, R. (2019). Germ granules in *Drosophila*. *Traffic (Copenhagen, Denmark), 20*(9), 650–660. https://doi.org/10.1111/tra.12674.

Valentin, G., Haas, P., & Gilmour, D. (2007). The chemokine SDF1a coordinates tissue migration through the spatially restricted activation of Cxcr7 and Cxcr4b. *Current Biology, 17*(12), https://doi.org/10.1016/j.cub.2007.05.020.

Venhuizen, J.-H., & Zegers, M. M. (2017). Making heads or tails of it: Cell–cell adhesion in cellular and supracellular polarity in collective migration. *Cold Spring Harbor Perspectives in Biology, 9*(11), a027854. https://doi.org/10.1101/cshperspect.a027854.

Venkiteswaran, G., Lewellis, S. W., Wang, J., Reynolds, E., Nicholson, C., & Knaut, H. (2013). Generation and dynamics of an endogenous, self-generated signaling gradient across a migrating tissue. *Cell, 155*(3), 674–687. https://doi.org/10.1016/j.cell.2013.09.046.

Wang, H., Guo, X., Wang, X., Wang, X., & Chen, J. (2020). Supracellular actomyosin mediates cell-cell communication and shapes collective migratory morphology. *iScience, 23*(6), https://doi.org/10.1016/j.isci.2020.101204.

Wang, X., He, L., Wu, Y. I., Hahn, K. M., & Montell, D. J. (2010). Light-mediated activation reveals a key role for Rac in collective guidance of cell movement in vivo. *Nature Cell Biology, 12*(6), https://doi.org/10.1038/ncb2061.

Weidinger, G., Wolke, U., Köprunner, M., Klinger, M., & Raz, E. (1999). Identification of tissues and patterning events required for distinct steps in early migration of zebrafish primordial germ cells. *Development (Cambridge, England), 126*(23), 5295–5307. https://doi.org/10.1242/dev.126.23.5295.

Wolf, K., & Friedl, P. (2006). Molecular mechanisms of cancer cell invasion and plasticity. *The British Journal of Dermatology, 154*(1), https://doi.org/10.1111/j.1365-2133.2006.07231.x.

Wolf, K., Mazo, I., Leung, H., Engelke, K., Von Andrian, U. H., Deryugina, E. I., ... Friedl, P. (2003). Compensation mechanism in tumor cell migration. *The Journal of Cell Biology, 160*(2), 267–277. https://doi.org/10.1083/jcb.200209006.

Wu, X., Tanwar, P. S., & Raftery, L. A. (2008). Drosophila follicle cells: Morphogenesis in an eggshell. *Seminars in Cell and Developmental Biology, 19*(3), https://doi.org/10.1016/j.semcdb.2008.01.004.

Yamaguchi, N., Zhang, Z., Schneider, T., Wang, B., Panozzo, D., & Knaut, H. (2022). Rear traction forces drive adherent tissue migration in vivo. *Nature Cell Biology, 24*(2), https://doi.org/10.1038/s41556-022-00844-9.

Yoon, C., Kawakami, K., & Hopkins, N. (1997). Zebrafish *vasa* homologue RNA is localized to the cleavage planes of 2- and 4-cell-stage embryos and is expressed in the primordial germ cells. *Development (Cambridge, England), 124*(16), 3157–3165. https://doi.org/10.1242/dev.124.16.3157.

Yoon, J., Hwang, Y.-S., Lee, M., Sun, J., Cho, H. J., Knapik, L., & Daar, I. O. (2018). TBC1d24-ephrinB2 interaction regulates contact inhibition of locomotion in neural crest cell migration. *Nature Communications, 9*(1), 3491. https://doi.org/10.1038/s41467-018-05924-9.

Zhang, N., Zhang, J., Purcell, K. J., Cheng, Y., & Howard, K. (1997). The Drosophila protein Wunen repels migrating germ cells. *Nature, 385*(6611), 64–67. https://doi.org/10.1038/385064a0.